GEOBOTANY AND BIOGEOCHEMISTRY IN MINERAL EXPLORATION

HARPER'S GEOSCIENCE SERIES
CAREY CRONEIS, EDITOR

GEOBOTANY AND BIOGEOCHEMISTRY IN MINERAL EXPLORATION

R. R. BROOKS

MASSEY UNIVERSITY
PALMERSTON NORTH, NEW ZEALAND

HARPER & ROW, PUBLISHERS
NEW YORK   EVANSTON   SAN FRANCISCO   LONDON

CONTENTS

PREFACE

The use of vegetation as a guide to mineralization is a field of research which is essentially multidisciplinary in scope and which involves such diverse subjects as biochemistry, biogeochemistry, biogeography, botany, chemistry, geobotany, geochemistry, geology, soil science, and statistics. These multidisciplinary aspects have tended to discourage research in the field since today is the age of specialization where workers skilled in "general science" are becoming increasingly rarer.

In this book, I have made an attempt to bring together the essentials of the subject within the compass of a single volume in the hope that this will be of assistance to those who wish to obtain the overall picture. With few exceptions, therefore, I have emphasized simplicity rather than detail. At the end of each chapter selected references are given for the benefit of those who wish to read more deeply into the subject discussed.

This work is addressed to several classes of reader. It is first of all designed for the field worker in exploration companies, who may be unacquainted with the subject and may wish to apply geobotany or biogeochemistry in his own areas. The book is also addressed to those students of biogeography or applied geochemistry who have yet to familiarize themselves with the application of their subject to the use of vegetation in mineral exploration. Other readers would include economic geologists, geologists, geophysicists, chemists, and biochemists, and the multitude of other workers whose disciplines overlap into this particular field.

In making acknowledgments, I often remind myself of the well-known quotation of the seventeenth-century English poet John Donne (1623), who wrote: "No man is an iland intire of it selfe." The essential truth of this maxim has never been more apparent to me than in the preparation of this book, because so much of what it contains I have learned from others. Above all I would like to acknowledge my debt to my present and former students Graeme Lyon, David Nicolas, Jeppe Nielsen, Barry Severne, Michael Timperley, and Neil Whitehead, all of whom have contributed something new to the field and to my own knowledge. I thank also my personal assistant Bertram Quin and former students Colin Boswell and Noel Cohen, who, though working in other fields, contributed expertise of direct benefit to geobotany and biogeochemistry.

I am grateful also to Professor R. D. Batt of Massey University for his encouragement in the establishment of a research group in this field at this university and for the provision of adequate facilities for research. Thanks are also due the Mineral Resources Sub-Committee of the New Zealand Universities Grants Committee for extensive and regular financial assistance during the past seven years, without which this work could never have been attempted. I would also like to thank my colleagues for criticism of various parts of the book, and I am grateful to my wife and Jean Thompson for assistance with typing.

Mention should also be made of financial and logistic support to my group from many exploration and mining companies in Australia and New Zealand. These include Australian Selection (Pty.) Ltd., Carpentaria Exploration Co. Pty. Ltd., The Consolidated Silver Mining Co. of New Zealand, Kennecott Explorations (Australia) Pty. Ltd., Lime and Marble Ltd., McIntyre Mines (N.Z.) Ltd., Norpac Mining Ltd., and Western Mining Corp. Generous support from these companies has much facilitated operations. Thanks are also due Mrs. Margaret Gwynn for assistance in the literature survey.

The foundation of this book is the pioneering work of many scientists such as Helen Cannon and Hansford Shacklette of the U.S. Geological Survey, Harry Warren and Robert Delavault of the University of British Columbia, and D. P. Malyuga of the former Vernadsky Institute of Moscow. Other workers, too numerous to mention, have also been a source of useful background information in the compilation of this work.

The stated aim of this book is to attempt to bridge the gap between the various disciplines involved in the study of plants as a guide to mineralization. If I have succeeded in persuading the geologist to drop his hammer for a moment and contemplate seriously the vegetation around him, or if I have succeeded in making the analyst and field worker more aware of their mutual problems, then I will be well satisfied.

R. R. Brooks

GEOBOTANY AND BIOGEOCHEMISTRY IN MINERAL EXPLORATION

CHAPTER ONE
GENERAL
INTRODUCTION

The scale of mineral exploration in the world is now of unprecedented proportions because of the increasing demands placed upon this planet's dwindling resources by the urgent needs of expanding economies and burgeoning populations. Most of the world's easily discoverable mineral deposits have already been investigated or exploited, and therefore it is only below the surface that major finds are likely to occur in the future. The pace of exploration is such that in New Zealand alone there are very few areas of geologically favorable ground for which prospecting licenses have not already been issued.

The search for minerals has been revolutionized not only by methods of exploration but also by the ancillary fields of chemical analysis and statistical interpretation of data. Among exploration methods there have been significant advances in geophysics as well as a steady refinement in geochemical procedures occasioned largely by progress in analytical chemistry.

Fortescue and Hornbrook (1967) have discussed the concept of a *prospecting prism* involving the different components of a landscape system which may be involved in mineral exploration. The prospecting prism shown in Figure 1-1 involves the concept of area as well as depth. From the figure it will be noted that "plant geochemistry" is included under the general heading of "geochemical methods of exploration" and is usually included in books and general reviews of the subject (e.g., Hawkes and Webb, 1962).

The use of vegetation as a guide to mineralization is considerably more complex than for example soil geochemistry, since it not only involves the response of plants to their environment but also includes two fields of study which are very different in scope and application. These two fields are *biogeochemistry* and *geobotany*. Biogeochemical methods of exploration depend on the chemical analysis of plants or humus to obtain evidence of mineralization in the substrate, whereas geobotanical methods involve a *visual* survey of the vegetation cover in order to detect mineralization by means of plant distributions, the presence of indicator plants,

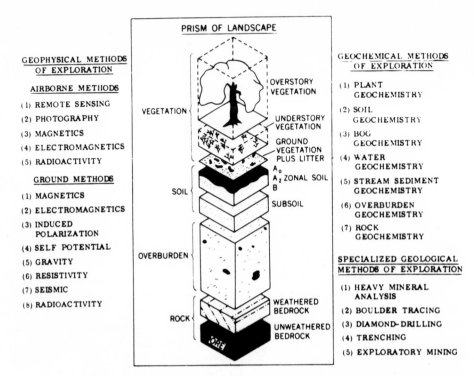

FIGURE 1-1 Diagram of a generalized prospecting prism showing the different components of a landscape that may be involved in prospecting methods. From Fortescue and Hornbrook (1967). By courtesy of the Geological Survey of Canada.

or mutational or morphological changes induced by excesses of certain elements in the substrate.

Geobotanical methods of exploration have been used for many centuries, whereas biogeochemical techniques have only been used since the late 1930s since they depend on the existence of speedy and reliable methods of analysis. World War II was followed by a boom in biogeochemical, geobotanical, and geochemical methods of exploration, particularly in the 1950s. Sanders (1967) has quoted an economic geologist at Harvard as saying:

> At the time [i.e., 1950–1960] some overzealous types looked upon geochemical prospecting as a panacea. We now realize all too clearly that geochemical methods, to be of genuine value in discovering ore deposits, must be combined with other methods—geophysical methods, geological reasoning, and so on. What we most certainly need is the total picture. . . .

This criticism is to some extent valid, but now that the initial euphoria has evaporated, it is possible to place geochemical procedures (including vegetation work) in proper perspective and to appreciate how valuable these methods are in the search for minerals when used in the correct manner, in the right place, and by the right people.

The ready acceptance of rock, soil, stream sediment, and water sampling as exploratory tools has not been extended to the same degree to biogeochemical or geobotanical methods of exploration. This is probably partly because such methods demand a greater degree of skill in execution and interpretation and partly because they usually require some sort of orientation survey in new areas. To some field workers, vegetation is a "nuisance," an impediment to their work which must be ruthlessly bulldozed out of the way before other operations can proceed. This attitude is fortunately changing, and many workers in the field of mineral exploration are beginning to realize that vegetation can be an asset rather than a hindrance. The distribution of species and the elemental content of the plant cover present a complex picture which can provide important information to the prospector if he has the knowledge and motivation to try to interpret it.

The extent to which plants may be used in the search for minerals will depend, of course, on the nature and extent of the vegetation cover. Draeger and Lauer (1967) have observed that two-thirds of the world's land surface is covered with vegetation, 42% of which comprises forest, 24% is grasslands, and 21% consists of desert shrubs and grasses in semiarid terrain.

Since developed areas with less natural vegetation will already have been explored more thoroughly than remoter regions, and because desert regions will have obvious outcrops, it is likely that most of the world's remaining mineral resources will be hidden beneath vegetation. For this reason alone, there should be increased scope for the future use of biogeochemistry and geobotany in mineral exploration.

1-1   **RESEARCH IN BIOGEOCHEMICAL AND GEOBOTANICAL METHODS OF EXPLORATION**

The progress of biogeochemistry and geobotany in mineral exploration is not easy to gauge since most of the work in this field is undertaken in the Soviet Union and is published in journals which are not always abstracted in the West. One criterion of progress is the annual number of entries in Chemical Abstracts, but this again is not completely reliable because the multidisciplinary nature of the subject renders abstraction by any single organization less likely. Despite these reservations, an attempt has been made in Figure 1-2 to give an estimate of the number of papers published in the two fields during the period 1938 to 1968. The information is based on the author's own card index which in turn is based on Chemical Abstracts, Biological Abstracts, and the bibliographies of published papers.

Table 1-1 lists the totals of all papers published in the period 1857 to 1968 and expresses the data on a regional basis. A distinction is made between biogeochemistry and geobotany and all papers have been assigned to either group depending on which subject makes the major contribution. In deciding whether or not to include ecological papers under geobotany, the deciding factor was that the aim of the work was to characterize geologic formations or mineralization within them.

Examination of Figure 1-2 shows a number of interesting features. Peaks of activity for Canada in the years 1947 to 1956 are almost entirely due to the work of

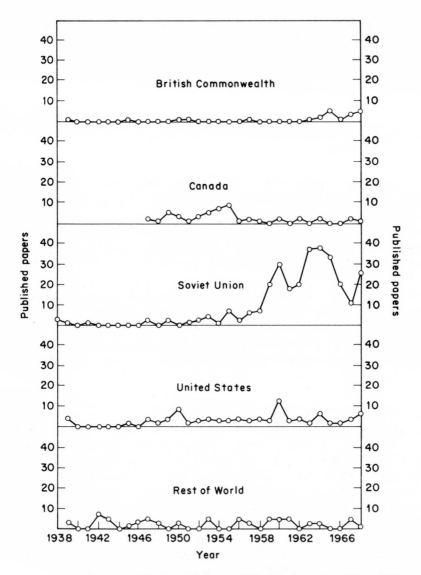

FIGURE 1-2 Numbers of papers published in the period 1938 to 1968 for the fields of geobotany and biogeochemistry in mineral exploration.

Warren and his co-workers, and indeed the level of activity during this period was at least as great as in the Soviet Union. An active peak for the rest of the world between 1939 and 1947 is largely due to the work of the Scandinavians, particularly Vogt. Most of the work in the United States for the period 1950 to 1964 was due to the efforts of Cannon, Shacklette, and other scientists of the U.S. Geological Survey. The figure shows the preponderance of the Soviet Union in this field. This can also be seen from Table 1-1, which shows that more than half of the world's

TABLE 1-1   Published Papers on Biogeochemistry (B) and Geobotany (G) in
Mineral Exploration for the Period 1857 to 1968

Region	Number of papers			Percentage of world total		
	B	G	B + G	B	G	B + G
British Commonwealth	18	6	24	5	3	5
Canada	42	4	46	13	2	9
Soviet Union	206	89	295	62	44	55
United States	35	61	96	11	31	18
Rest of World	31	39	70	9	20	13
Total	332	199	531	—	—	—

effort has been concentrated in that country. The dominance of the Soviet Union in biogeochemistry and geobotany has been even more pronounced during the 1959 to 1968 decade, when 73% of the papers published in these fields were Russian. This figure compares with 13% for the United States, 5% for the British Commonwealth, 2% for Canada, and 7% for the rest of the world.

The data in the above table and figure also show that biogeochemistry is used more often than geobotany in mineral exploration, particularly in the Soviet Union. It must also be emphasized that the figures for the Soviet Union are somewhat speculative for the reasons enumerated in the first paragraph of this subsection. Compilations up to 1959 are based largely on the bibliography in Malyuga (1964); but since about one third of these Russian references were not found in Chemical Abstracts, the number of Russian references after 1959 was calculated by taking the entries in Chemical Abstracts and increasing the numbers by 50%.

1-2   CENTERS OF RESEARCH

There are relatively few institutions in the world where active research into biogeochemical and geobotanical methods of exploration is presently undertaken. Table 1-2 lists such institutions; as will be noted, the Soviet Union is again predominant. The total is indeed incomplete since it does not include several institutions within the Academies of Science of other Soviet Republics where work of this nature is also carried out.

Table 1-3 lists individual workers who have published more than five papers in the two fields. The table also gives the data of the first publication and that of the last, if the scientist is known to have ceased working in the field. As before, the numbers quoted for the Russian workers may be conservative.

**1-3   THE LITERATURE OF BIOGEOCHEMISTRY AND
GEOBOTANY IN MINERAL EXPLORATION**

Papers concerning plants as a guide to mineralization are distributed over a very wide range of journals because of the interdisciplinary nature of the subject. Table 1-4 gives a compilation of the number of publications in this field found in the journals most commonly used to report the data and does not include any marginal papers.

TABLE 1-2   Some Centers of Research into Biogeochemical and Geobotanical Methods of Mineral Exploration

Country	Institution	Address
Canada	Geological Survey of Canada	Ottawa, Ontario
	Geology Department, University of British Columbia	Vancouver, British Columbia
New Zealand	Department of Chemistry and Biochemistry, Massey University	Palmerston North, New Zealand
Soviet Union[a]	All-Union Aerogeologic Trust (VAGT)	Moscow
	All-Union Scientific Research Institute of Exploration Geophysics (VIRG)	Moscow
	Buryat Interscience Research Institute	Ulan Ude, USSR
	Department of Geography, Moscow State University	Moscow
	Institute of Geology and Geophysics	Tashkent
	Vernadsky Institute of Geochemistry and Analytical Chemistry	Moscow
United Kingdom	Department of Geography, Bedford College	London, N.W.1
United States	U.S. Geological Survey	Denver, Colorado

[a]There are several other institutes of the Academies of Science of other Russian Republics also undertaking this type of work.

TABLE 1-3   Workers Who Have Published More Than Five Papers in the Fields of Biogeochemistry and Geobotany in Mineral Exploration

Name	Number of papers	Period	Institution
Brooks, R. R.	16	1966–	Massey Univ., Palmerston North, New Zealand
Cannon, H. L.	28	1951–	U.S. Geological Survey, Denver
Delavault, R. E.	23	1948–	Dept. Geology, Univ. British Columbia, Vancouver
Ernst, W.	13	1964–	Dept. Applied Botany, Wilhelms Univ., Muenster, Germany
Kovalevsky, A. L.	15	1960–	Buryat Interscience Res. Inst., Ulan Ude, USSR
Koval'sky, V. V.	11	1958–	Vernadsky Inst., Moscow
Malyuga, D. P.	17	1947–	Vernadsky Inst., Moscow
Robinson, W. O.	9	1945–1958	—
Talipov, R. M.	8	1964–	Inst. Geol. Geophys. im. Abdullaeva, Tashkent, USSR
Tkalich, S. M.	6	1938–1959	—
Shacklette, H. T.	18	1958–	U.S. Geological Survey, Denver
Viktorov, S. V.	10	1955–	Dept. Geography, Moscow State Univ.
Vogt, T.	8	1939–1948	—
Warren, H. V.	27	1947–	Dept. Geology, Univ. British Columbia, Vancouver

Several reviews and books on the subject have appeared in the past decade. These include works by Brooks (1968), Cannon (1960b), Carlisle and Cleveland (1958), Chikishev (1965), Fortescue and Hornbrook (1966, 1967), Malyuga (1964), Nesvetaylova (1961), Pauli (1968), Rommel et al. (1968), Thaler (1962), Usik (1969), and Viktorov et al. (1964).

TABLE 1-4   Journals with More Than Five Biogeochemical and
Geobotanical Contributions to the End of 1969

Name of journal	Number of contributions
Geokhimiya	18
Publications of U.S. Geological Survey	16
Economic Geology	15
International Geology Reviews	10
Kgl. Norske Videnskab. Selskab Forh.	9[a]
Bull. Geological Society of America	8
Trudy Biogeokhimichesky Lab.	8
Trans. Royal Society of Canada	7
Soil Science	6
Publications of Geological Survey of Canada	6
Botanichesky Zhurnal	6

[a] All contributions from Vogt and his co-workers.

1-4   THE SCOPE AND AIMS OF THIS BOOK
The purpose of the present work is to describe the nature and potential of biogeochemical and geobotanical methods of prospecting and to present the essentials of the wide range of disciplines that are associated with it. No extravagant claims will be presented for these methods, but an attempt will be made to present the facts before the reader in order that he may fully realize the extent to which fitting the final piece to the jigsaw of the prospecting prism will serve to present a clearer picture than any one method alone.

PART ONE
GEOBOTANY
IN
MINERAL
EXPLORATION

CHAPTER TWO
INTRODUCTION TO GEOBOTANY IN MINERAL EXPLORATION

Geobotanical methods of prospecting involve the use of vegetation for identification of the nature and properties of the substrate. Paradoxically, these methods are among the easiest to execute and yet the most difficult to interpret of all the methods of exploration available at the present time. In terms of execution, the basic requirement is merely a pair of human eyes; but in the interpretation of the visual (or photographic) image, some knowledge is required of a number of different disciplines such as biochemistry, botany, chemistry, ecology, geology, and plant physiology. Since there are very few people with a good working knowledge of all of these different subjects, it is often necessary to resort to the use of teams of experts in order to develop the method to its full potential.

The success of geobotanical methods in the future will depend to a large extent on the degree to which man can be trained to interpret the information obtained from a study of the vegetation cover. In the words of the sixteenth-century physician, Andrew Boorde (Breviary of Helthe, 1547): "Who is blinder than he who will not see?"

In geobotanical methods of prospecting, the interpretation of the significance of the vegetation cover is carried out in a number of ways. These are: studying the nature and distribution of plant communities; studying the nature and distribution of indicator plants; observation of morphological changes in vegetation (e.g., color changes); examination of any of the above factors by aerovisual observations or by aerial photography.

The use of plant communities or indicator plants in the search for water or minerals has been known since very early times. Even in Rome in the reign of Augustus Caesar (63 BC to 14 AD), the famous architect Vitruvius wrote:

> The slender bulrush, the wild willow, the alder, the agnus castus, reeds, ivy and the like . . . usually grow in marshy places . . . but water is to be sought in those regions and soils, other than marshes, in which such trees are found naturally and not artificially planted.

Most of the geobotanical observations recorded in medieval and Roman times

(Granger, 1962) were concerned with hydrological data involving the search for water in arid regions. Even today this is still a major part of the effort being applied in various parts of the world. Cannon (1960b), Thaler (1962), and Viktorov (1955) have reviewed the history and present status of the geobotanical method. Chikishev (1965) has given an extensive review of plant indicators of soils, rocks, and subsurface waters.

Since World War II, aerial methods of observation (visual and photographic) have played an increasingly important role in the science of geobotany. In the Soviet Union, the All-Union Aerogeologic Trust (VAGT) was established in 1945 in order to carry out geobotanical surveys in connection with such diverse fields as agriculture, geologic mapping, hydrology, mineral exploration, and silviculture. Similar work, though on a much smaller scale and with less coordination, has been carried out in North America and in Australia.

In the following chapters of Part I of this volume, the broad outlines of geobotanical methods of prospecting are presented and discussed. As far as possible, an attempt has been made to give some theoretical basis of the concepts presented without delving too deeply into biochemistry or plant physiology. It is the author's considered opinion that a geobotanist who has some concept of the theoretical basis of his work will, in the end, prove to be more effective than one who carries out his work on a purely empirical basis without regard to the principles involved.

In discussing the subject of geobotany in mineral exploration, an attempt has been made to avoid the danger of overstating the potentiality of the method. It is not claimed that geobotany itself will of necessity give the whole picture of mineralization in a given area but it is certainly claimed that the technique is of great value in completing the picture and in some cases can decisively represent the difference in success or failure of an exploration project. The procedure is above all dependent on the skill of the operator; and unless skilled personnel can be employed, disenchantment will follow.

CHAPTER THREE
PLANT COMMUNITIES IN INDICATOR GEOBOTANY

The Russian geologist Karpinsky (1841) was one of the first to recognize that different plant associations exist on varying geologic substrates such as sandstones, clays, limestones, etc., and that this might be used to characterize the geology of the area concerned. Karpinsky concluded that reliance should be placed on an examination of the whole community rather than on one or two characteristic species within it. His classical work has ultimately led to the development of the science of *indicator geobotany* which is a division of geobotany concerned with the theoretical and practical aspects of vegetation and its component species as indicators of the conditions of the environment.

Indicator geobotany has been brought to a very high state of development within the Soviet Union, where an extensive literature exists on this subject. Among the best reviews to appear in recent years are those of Chikishev (1965) and Viktorov et al. (1964).

Indicator communities or *characteristic floras* will not in themselves necessarily indicate mineralization but will often serve to characterize regions where certain types of mineralization are likely to occur. Examples of this are the use of serpentine floras for locating chromite deposits (Lyon et al., 1968) or the study of selenium floras (Cannon, 1957), which can indirectly show the presence of uranium mineralization since this element is frequently associated with selenium. Linstow (1929), in reviewing the field, has referred to such communities as *bodenanzeigende Pflanzen* (soil-indicating plants).

Characteristic floras can be useful in field work whenever geologic maps are either lacking or are of poor quality. If geobotanical mapping is carried out in the area, a reliable indication can often be obtained of the nature of the substrate and the delineation of its boundary with other geologic formations.

Malyuga (1964), in discussing geobotanical mapping of plant associations, has observed that it is of limited use unless it is combined with other geochemical or biogeochemical investigations. Figure 3-1 shows a comparison of schematic geo-

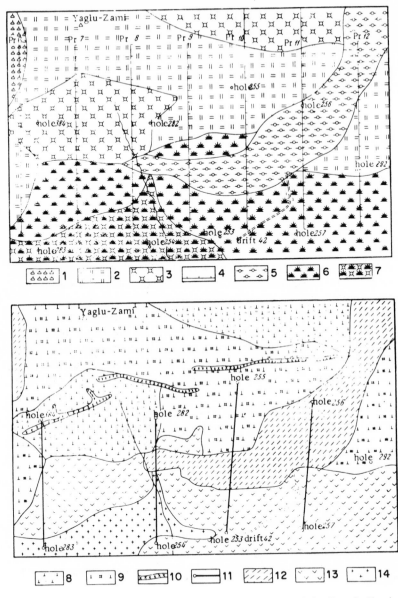

FIGURE 3-1 Geobotanical and geologic maps of the Karmir–Karsky area of the Soviet Union. *Upper map*: 1, *Silene compacta* association; 2, grass association; 3, legume–grass association; 4, biogeochemical survey profiles; 5, *Lapsana communis* association; 6, thyme–tragacanth association; 7, thyme–tragacanth association with legumes and various grasses. *Lower map*: 8, weakly modified porphyrites; 9, strongly modified porphyrites; 10, granodiorite–porphyry dikes; 11, geologic profiles between holes; 12, hornfels; 13, unmodified monzonites; 14, syenites. From Malyuga (1964). By courtesy of the Plenum Publishing Corp.

logic and geobotanical maps of the Karmir–Karsky area near Kadzharan in Soviet Armenia where it can readily be seen that there is a striking correlation between the geology and plant associations of this region.

3-1   MAPPING TECHNIQUES IN INDICATOR GEOBOTANY

In many cases the distinction between two different plant communities is so pronounced that mere visual observation is all that is needed to delineate the geologic boundary between their substrates. This is particularly the case for serpentine floras (Cuyler, 1931). Even when the rock types are relatively similar, pronounced differences in the vegetation can occur. This has been demonstrated by Billings (1950) for vegetation growing over altered and nonaltered andesite in Nevada.

In cases where superficial observation is not sufficient, recourse must be made to geobotanical mapping. Such an operation should ideally be carried out by a skilled botanist or ecologist since the amateur can readily confuse plants which are superficially similar and would have difficulty distinguishing between different *ecotypes* of the same species even if they were in fact distinguishable.

Geobotanical mapping involves the selection of a number of sample plots known as *quadrats.* There is no general agreement on the best method of selection of quadrats. Some workers consider that they should be selected in a random manner whereas others believe that they should be chosen subjectively. This latter procedure, though frequently used, is open to some criticism insofar as it makes the assumption that the vegetation associations are already known.

If the subjective approach is to be used, the following procedure should be employed. Every plot should have the utmost uniformity it is possible to find in the area concerned; not only with regard to plant species, but also with consideration of such factors as aspect, slope, drainage, relief, and altitude. Particular care must be taken that the quadrats do not include two or more different associations as may occur, for example, with plots not of uniform slope.

The size of the quadrat now has to be established. As a general rule the size should be the minimum needed to include most of the plants of the association and will obviously be a function of the homogeneity of the community.

In assessing the size of the *minimal area*, the law of diminishing returns will obviously apply; successive increases in size of the sample area will give successively smaller amounts of additional information. The concept of minimal area has been throughly reviewed by Goodall (1952). A species-area plot at first rises sharply and then becomes flatter, although it never becomes completely horizontal because the whole area would have to be included in the quadrat in order to incorporate every single species present. Greig-Smith (1964) has discussed the problem of minimal area at some length, and there appears to be no general agreement on a universal criterion to determine this area. As a general rule however, the following procedure may be adopted in the field.

Begin with a small quadrat of perhaps 5 sq m, note the species within it, and then increase the size of the plot progressively (10, 50, 100 sq m, etc.) noting at each stage additional species included. When there is an appreciable drop in the rate of increase of new species found, the optimum size of quadrat will have been found.

When the size and positions of the test plots have been established, the next procedure will involve an evaluation of the *density* of individual species and their *spacing* (reciprocal of density). In determining density, direct counting or a scale of numbers may be used. The scale is somewhat arbitrary but can give good results in the hands of an experienced operator. The system is as follows: 1, very rare; 2, rare; 3, infrequent; 4, abundant; 5, very abundant. One disadvantage of this system is that data are heavily dependent on the personal assessment of one individual and are not always comparable with data collected by other workers. The use of this scale is nevertheless justified on the grounds of speed and practicality.

A geobotanical map of an area need contain nothing more than the density or spacing data enumerated above. If other parameters are added, a more meaningful map can be compiled.

The space demand of a species involves another concept, that of *cover*. It is assumed that the entire shoot system of the plant is projected on the ground and that this area (equal to the area of shade if the sun is directly overhead) represents the cover. Cover is usually expressed as a percentage. The total for all species will often exceed 100% due to overlap. Methods of determining cover have been reviewed by Brown (1954).

One quick method of measuring cover is to determine the total length of interceptions made by plants on line transects. The proportion of the total length of the transect intercepted by a species gives a measure of its cover. For other methods of measurement, the reader is referred to Greig-Smith (1964).

Some mention should be made of the concept of *layering*. The vegetation may be considered not only laterally but also vertically. In the vertical concept, a number of distinct layers are recognized. These are the tree layer, shrub layer, herb layer, and moss layer. Clearly, the denser the upper tree layer, the greater will have to be the tolerance toward reduced light intensity by the lower layers of the community. Many mosses, as might be expected, will tolerate the least light intensity.

There are a number of other criteria of plant communities which can also be employed in geobotanical mapping. These are, however, somewhat outside the scope of this book and will only be mentioned briefly.

Sociability expresses the space relationship of individual plants and can be expressed in terms of a simple scale (Braun-Blanquet, 1932) as follows: soc. 1, growing on one place singly; soc. 2, grouped or tufted; soc. 3, in troops, small patches, or cushions; soc. 4, in small colonies, in extensive patches, or forming carpets; soc. 5, in great crowds.

Vitality is a measure of how a plant prospers in a given area and can be expressed by a number of conventional symbols (Braun-Blanquet, 1932) as follows. ●—well-developed, regularly completing the life cycle; ☉—strong and increasing but usually not completing the life cycle; ◎—feeble but spreading, never completing life cycle; ○—occasionally germinating but not increasing, many ephemeral plants.

Periodicity is a measure of the regularity or absence of rhythmic phenomena in plants, such as flowering, fruiting, etc. A study of this criterion involves continuous and systematic research and is outside the scope of this work.

If mean values for density, spacing, cover, and other parameters are computed for each species and averaged over several quadrats (preferably chosen randomly), it will be possible to characterize the plant associations in the test area and to attempt to correlate this with the geologic environment.

Although the use of quadrats is the most usual approach in indicator geobotany, a different approach will be needed for studying plants growing over narrow ore bodies. In such cases the use of *line transects* or *belt transects* is recommended. Line transects consist of parallel straight lines run through an area with the aid of tape measures and a compass. The length of each line through each community is recorded. Belt transects consist of a continuous series of quadrats running across the profile of the area. Figure 3-2 shows data for a belt transect in a region of Western Australia.

When the basic data have been obtained for a geobotanical map, the next problem is to decide how representation should be carried out cartographically. There

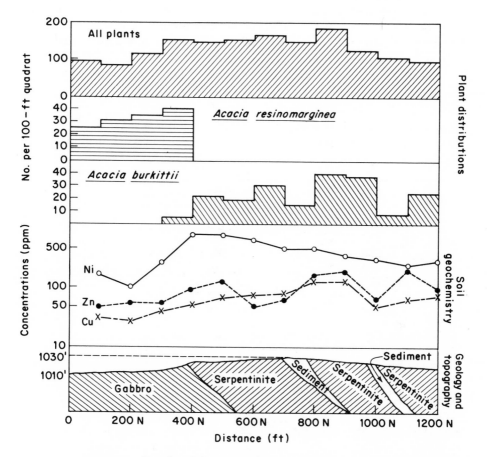

FIGURE 3-2 Data from belt transect of an area in Western Australia showing the apparent influence of the geology, topography, and soil geochemistry on the distribution of two plant species. By courtesy of Western Mining Corporation Ltd.

are no international standards for this purpose, although in the Soviet Union a relatively standard system of colors is used to represent various plant communities (Viktorov et al., 1964). As an alternative to colors, the Baranov system (Baranov, 1933) may be used. In the Baranov system, a series of symbols (circles, lines, arrows, etc.) are combined together to represent the entire plant community and the abundance of its constituents. The combined symbols are used on the appropriate areas on the map.

For further information on geobotanical mapping the reader is referred to Braun-Blanquet (1932) and Viktorov et al. (1964). For other methods of representing geobotanical data, reference should be made to Cole et al. (1968), Greig-Smith (1964), and Nicolls et al. (1965).

3-2   CHARACTERISTIC FLORAS

The ecology of a plant community will be greatly influenced by the pH of the soil and by the presence, excess, or deficiency of mineral nutrients. The *availability* of elements is affected by the pH of the substrate, as can be seen from Figure 3-3 which represents the relative availability of a number of different elements at various pH values. The pH of the soil can also exert an influence insofar as oxidation potentials for most reactions are pH dependent. For example in the reaction:

$$Fe^{3+} + e = Fe^{2+}$$

the oxidation potential (IUPAC convention) is $+0.6\ V$ at a pH of 4 and is about $-0.2\ V$ at pH 8. For this reason, iron will tend to be oxidized at high pH values,

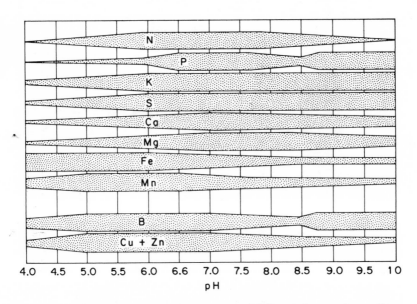

FIGURE 3-3 Effect of pH on the availability of plant nutrients. From Ignatieff (1952).

in which state it becomes less available to plants than in the reduced form which is more stable at a low pH.

It is obvious from the above discussion that the Eh of the soil will also affect availability by controlling the oxidation state of an element at a particular pH value.

If certain elements or groups of elements predominate in the substrate, they can affect the ecological balance either directly by their toxic effects or indirectly by their *antagonistic effects* on other elements. At certain concentrations, these elements will be toxic to most plants and only those species able to adapt to the hostile environment will survive.

Figure 3-4 shows the antagonistic and stimulatory effects of various ions. For example, high copper or zinc levels in the soil should depress iron uptake and produce chlorosis in the vegetation.

The influence of nutrients on the ecology of plants is neatly summarized by the *Law of Relativity* (Lundegårdh, 1931) which may be expressed as follows: "The more nearly a factor is in minimum in relation to the other factors acting upon the organism, the greater is the relative influence of a change of that factor upon the growth of the organism." In other words, as a factor increases in intensity, its relative effect on the organism decreases; and when the factor is in the region of its maximum, the effect of a change on the organism is nil.

It would no doubt be possible to find characteristic floras for nearly all soil

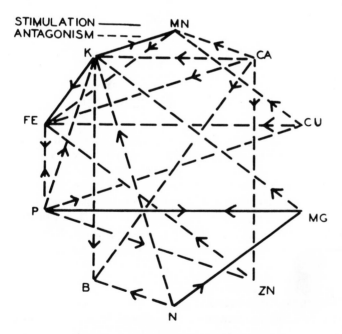

FIGURE 3-4 Stimulatory and antagonistic effects of various pairs of elements on their absorption by plants. From Mulder (1953).

TABLE 3-1   Characteristic Floras

Name of community type	Locality	Substrate	References
Calciphilous flora	Harz Mts.; Derbyshire; S. Tyrol	Gypsum, limestone, dolomite	Ellenberg (1958)
Halophytic flora	Utah, Colorado; Steppes of USSR; Australian deserts	Sodium chloride, sodium carbonate, gypsum	Adriani (1958) Braun-Blanquet (1932) Simmoneau (1954)
Selenium flora	Colorado; Alberta; Colombia; Queensland	Seleniferous sandstones	Beath et al. (1935) Cannon (1952, 1957, 1960a, 1964) Miller and Byers (1937) Peterson and Butler (1967) Trelease and Beath (1949)
Serpentine flora	Nelson Province, New Zealand; Sweden; N. Finland	Serpentine and other ultramafics	Igoshina (1966) Lounamaa (1956) Lyon et al. (1968) Minguzzi and Vergnano (1953) Robinson et al. (1935) Rune (1953) Sarosiek (1964) Whittaker (1954)
Zinc ("Galmei") flora	Belgium; Germany	Zinc carbonate, zinc silicate	Baumeister (1954) Ernst (1967) Linstow (1929) Schwanitz and Hahn (1954a, 1954b) Schwickerath (1931)

types and geologic formations, but certain ecological groupings are well documented (these are shown in Table 3-1).   Each plant community will be discussed
briefly in an attempt to present to the reader the basic concepts underlying the
reasons for the existence of specialized floras. In the past much field work has been
carried out on a purely empirical basis with little consideration of the basic principles involved. A field worker with an adequate knowledge of why characteristic
floras exist should be all the more effective in apply geobotanical methods to
mineral prospecting.

3-3   CALCIPHILOUS (LIMESTONE) FLORAS
        Calcium has a great effect on vegetation. It is not surprising, therefore,
that calcareous rocks such as limestone and dolomite usually carry a characteristic
flora which often renders the geologic boundary easy to delineate. The effect of

calcium on plants depends more on its solubility than on the absolute amounts present. The solubility of limestone in water is about 0.1%, whereas dolomite is about half as soluble.

Many of the effects of calcium are indirect, as lime has a favorable effect on the drainage and structure of the soil since it is able to coagulate many of the colloidal compounds which make a soil heavy and poorly drained. Limestone soils are therefore well aerated and are good conductors of water and heat. The result of this conditioning effect is that many plants which thrive in any type of soil in warm and dry climates tend to confine themselves increasingly to calcareous soils at higher latitudes since these are the only types of soil that can provide them with the required physical conditions. A particularly striking example of this factor is the distribution of the beech *Fagus sylvatica* which thrives in Southern Europe on any type of soil whereas in England it tends to be found in its natural state mainly on chalk.

Plant communities which thrive on calcareous soils may be said to be *calciphilous*, benefiting either by the direct or indirect effect of calcium. Plants having a direct and specific requirement for calcium are said to be *calcicolous*. Characteristic floras for limestones therefore contain both types of community and exclude the *calcifuge* species (such as the *Ericaceae*) which require a low pH in order to thrive.

Calciphilous plants have been studied for well over a century (Karpinsky, 1841; Ellenberg, 1958; Chikishev, 1965). Because limestone does not produce a peculiar or stunted flora, this poses problems in the identification of the substrate. By its very richness, the flora presents a problem in geobotanical mapping. One way around this difficulty is to look for certain species known to be characteristically calciphilous such as the genera *Dianthus*, *Fagus*, *Bromus*, *Festuca*, *Linaria*, etc. Another method of approach is to compare the richness of a limestone community with the relative paucity of floras on other adjoining substrates.

3-4   HALOPHYTE FLORAS

A characteristic plant association is found over saline soils containing sodium chloride, sodium carbonate, or sodium sulfate. Saline soils are found usually in areas of dry climate or near the sea. The ecology of *halophytes* has been studied by several workers (Simmoneau, 1954; Adriani, 1958). Extensive discussion of this plant association is beyond the scope of this book since much of the work has been carried out on semimarine species such as mangroves. In geobotanical mapping, it is the halophyte communities of dry areas which have received the most attention.

Typical dry-climate halophyte associations are found in Australia (*Atriplex* spp.), the Western United States, and the steppe regions of the Soviet Union. Halophytes usually have a high osmotic pressure in the cells and have the ability to accumulate relatively large quantities of salts.

In the Soviet Union halophytes have been studied extensively as a guide to water resources in hydrogeochemical studies (Chikishev, 1965). An interesting subgroup of the halophytes is the selenium floras, which are discussed in the next section. These are found in conjunction with the *Atriplex* association in the Colorado Plateau (Cannon, 1957).

3-5   SELENIUM FLORAS

One of the most interesting cases of the use of a characteristic flora in the search for minerals has been the discovery of the selenium flora of the western United States (Beath et al., 1935, 1939a, 1939b), Colombia, Canada, and Queensland (McCray and Hurwood, 1963). Selenium communities indicate the presence of selenium in the soil either because they have a specific requirement for this element or because they can tolerate large concentrations of it.

Selenium indicators (Trelease and Beath, 1949) include certain species within the genera *Astragalus*, *Stanleya*, *Aster*, and *Oryzopsis* and are found among the characteristic halophytic "shadscale" (*Atriplex confertifolia*) associations. The Australian plants include *Neptunia amplexicaulis* and *Acacia cana*. Although selenium-indicating plants contain high concentrations of this element, they are probably capable of growth without selenium (Shrift, 1969). These species are apparently able to substitute selenium for sulfur in large sectors of their metabolism without toxic effects (Shrift, 1969).

Uranium deposits on the Colorado Plateau are primarily carnotite containing appreciable quantities of selenium. The presence of carnotite results in greater availability of selenium to plants. Cannon (1952, 1957, 1960a, 1964) has been able to use the *Astragalus* species for indirect prospecting for uranium because the plants tend to grow in areas of maximum total or available selenium in the soils. Her classic work of a decade ago represents one of the most successful known applications of the geobotanical method. This work will be discussed further in the next chapter.

Some of the members of a selenium flora are able to absorb such large quantities of this element (up to 1.5% of the ash weight of some *Astragalus* species) that the garliclike odor of volatile selenium compounds can often be detected in the plants themselves, sometimes even from a fast-moving automobile. It might be unreasonable to expect that this represents the ultimate in the geobotanical method in that the sense of smell can be used in conjunction with the sense of vision for the assessment of the mineral potential of an area.

The plant chemistry of selenium floras has been studied extensively by various workers (Horn and Jones, 1941; Trelease et al., 1960; Peterson and Butler, 1962; Virupaksha and Shrift, 1965; Shrift and Virupaksha, 1965; Peterson and Butler, 1967; Shrift, 1969). Investigations have shown that selenium accumulators differ from other species not only in the amount of uptake of selenium but also by the fact that this element occurs predominantly in nonprotein amino acids and it is in this way that a selenium flora has been able to adapt to high soil concentrations of this element.

FIGURE 3-5 Aerial photograph of the Dun Mountain ultrabasic area near Nelson, New Zealand. The characteristic serpentine flora is shown by the light-colored areas which are at a lower altitude than the surrounding forested region. Photograph Crown Copyright, by courtesy of the Lands and Survey Department of New Zealand.

3-6   SERPENTINE FLORAS

Of all morphological changes produced on vegetation by the substrate, those found in serpentine floras are perhaps the most extreme. So great is the differentiation between serpentine floras and the surrounding types of vegetation that boundaries between them are readily observed. Figure 3-5 shows an aerial photograph of the Dun Mountain (original type locality for the mineral dunite) ultramafic area near Nelson, New Zealand. The white areas designate the serpentine vegetation. Figure 3-6 shows a close-up photo of the boundary between the serpentine vegetation and the normal New Zealand bush growing over a sedimentary formation.

Extensive studies of serpentine floras have been carried out in a few areas: for example, New Zealand (Lyon et al., 1968; Lyon et al., 1970); Poland (Sarosiek,

FIGURE 3-6 View of Wooded Peak, Nelson, New Zealand, showing the sharp vegetation transition between the serpentine (foreground) and sedimentary formations (background).

1964); the Soviet Union (Igoshina, 1966); Sweden (Rune, 1953); and the United States (Robinson et al., 1935). Typical examples of this type of flora show a general sparseness of vegetation with a shortage of species as well as individuals. There are usually a few species which are endemic to a particular area, such as *Myosotis monroi* and *Pimelea suteri* in the Dun Mountain area of Nelson Province, New Zealand.

Serpentine soils are considerably different from normal soils, being rich in chromium, cobalt, iron, magnesium, and nickel but deficient in calcium, molybdenum, and the macronutrients nitrogen, phosphorus, and potassium.

Lounamaa (1956) and Robinson et al. (1935) concluded that the unusual serpentine flora results from excessive amounts of chromium, cobalt, and nickel. Other workers, such as Minguzzi and Vergnano (1953) and Rune (1953), considered that nickel was particularly responsible for the specialized vegetation.

The low calcium level of serpentine soils led Kruckeberg (1954), Walker (1954), and Walker et al. (1955) to conclude that the flora on such soils is unusually tolerant of a high magnesium : calcium ratio. Yet another theory to account for the existence of serpentine floras is that of Krause (1958), Paribok and Alekseeva-Popova (1966), and Sarosiek (1964), who suggested that the survival of plants on serpentine soils depends on their ability to adapt, at least partially, to all the factors of a serpentine soil which render it unfavorable to their development.

Recently, Lyon (1969); Lyon, Brooks, and Peterson (1969); Lyon et al. (1968); Lyon et al. (1970); and Lyon, Peterson, and Brooks (1969) have carried out an extensive investigation on a New Zealand serpentine flora and have come to the conclusion that some plants have an ability to extract calcium to a greater extent than other nonserpentine plants and in this way counteract to some extent the deficiency of calcium in ultramafic soils.   The same authors have confirmed that serpentine plants are able to accumulate extraordinary amounts of nickel and chromium.   The endemic New Zealand serpentine species *P. suteri* contains over 0.5% nickel and 0.2% chromium in its ash.

It is clear that the "serpentine problem" is far from solved and that much work remains to be done before it can be established with certainty why this peculiar ecological community exists in ultramafic areas.   It is probable that the main factors influencing the development of a serpentine flora vary in different areas and that no universal criteria exist.

3-7   ZINC ("GALMEI") FLORAS

Plant communities growing over soils high in copper, lead, or zinc have a certain similarity with serpentine floras: plant growth is retarded, broadleaf plants are absent, and endemic forms are to be found (Schwickerath, 1931; Robyns, 1932; Baumeister, 1954; Schwanitz and Hahn, 1954a). Some species such as *Agrostis tenuis*, *Silene cucubalus*, and *Campanula rotundifolia* produce ecotypes in these areas indistinguishable from those growing over serpentine soils (Prat, 1934; Bradshaw, 1952; Schwanitz and Hahn, 1954b).

In many cases it cannot be established whether or not the characteristic flora is due to one or all of the three base metals referred to above; they are often found together, particularly in areas of sulfide mineralization.   There has been a tendency to classify such plant communities as zinc (or galmei) floras since zinc is usually the main constituent.

The true zinc floras are found in Western Germany and Belgium where the soils are rich in zinc and do not contain inordinate amounts of copper or lead.   Galmei floras have been known for well over 100 years and early miners were guided to ore deposits by members of the community such as *Viola calaminaria*.   The capacity of galmei floras to accumulate zinc is quite remarkable and Linstow (1929) reports that *Thlaspi calaminare* contains ten times as much of this element in the leaves as in the roots.

An excellent review of the literature concerning zinc and other heavy-metal floras has been compiled by Ernst (1967), who lists ninety-six references.

3-8   OTHER CHARACTERISTIC FLORAS

A full discussion of other characteristic floras is beyond the scope of this work; for further information the reader is referred to Adriani (1958).   In many instances evidence for some characteristic floras is sparse and based on the distribution of so few species that the plants may better be considered as individual indicators (see Chapter 4) rather than part of a specific ecological community.

SELECTED REFERENCES

Indicator geobotany: Chikishev (1965); Malyuga (1964); Viktorov et al. (1964)

Geobotanical mapping: Gates (1949); Viktorov et al. (1964)

Calciphilous floras: Ellenberg (1958)

Halophytes: Adriani (1958)

Selenium floras: Cannon (1957); Trelease and Beath (1949)

Zinc floras: Ernst (1967); Schwanitz and Hahn (1954a, 1954b)

CHAPTER FOUR
INDICATOR
PLANTS

In the previous chapter consideration was given to plant communities indicative of the type of substrate upon which they grow. Such floras do not necessarily localize mineralization within the area concerned but are nevertheless useful aids in geobotanical mapping. This chapter is concerned with *indicator plants* which enable a mineralized occurrence to be located more precisely.

From the very earliest times it has been recognized that a number of plant species have a preference for certain types of mineralization and that this factor can be used for prospecting purposes. One of the earliest examples of this was the *copper plant* (*kisplant*), probably *Viscaria alpina*, which was used by medieval miners in Scandinavia in their search for copper and other pyrite ores (Vogt, 1942a).

Indicator plants are divided into two main classes according to their distribution. The first group comprises the so-called *universal indicators* which will not grow in nonmineralized substrates and can be used in any region in which they occur. These indicators are immensely valuable in prospecting since their presence almost always indicates a high soil concentration of the element being sought, but they do suffer from the disadvantage that they are usually rare and of a limited distribution range. In spite of these disadvantages, the greatest successes in geobotany have been obtained with their aid. Examples are the discovery of copper in Zambia by use of *Ocimum homblei* (now *Becium homblei*) (Horizon, 1959) and the use of *Astragalus* for discovering uranium in the Colorado Plateau (Cannon, 1957, 1960a, 1964).

Another group of plants comprise the *local indicators* which are species adapted to tolerating mineralized ground but which will grow elsewhere provided that competition from other species is not too great. Such indicators are usually considerably more common than universal indicators but have the disadvantage that they are often only useful in a limited area.

The use of plant indicators as a guide to mineralization is somewhat complicated by the fact that the element which controls the distribution of the plant may be either the one which is sought or an associated element or condition (e.g., pH or water availability). When the plant gives a *direct* response to the element, it may be said to be a *primary indicator*. When response is indirect, the species is known as a *secondary indicator*.

Table 4-1 lists a number of indicator plants and is based on an original table from Cannon (1960b). Only sixty-two species are listed in an attempt to restrict the names to only those plants known to have been used in prospecting. The current

TABLE 4-1   Indicator Plants

Element	Species [a]	Locality	Reference
Bitumen	*Artemisia limosa* (L)	Siberia	Popov (1949)
	Primula sachalinensis (L)	Siberia	Popov (1949)
	Gentiana paludicola (L)	Siberia	Popov (1949)
	Salsola rigida var. foliosa (U)	Siberia	Yaroshenko (1932)
	Astragalus aureus (L)	U.S.S.R.	Burenkov and Kuzina (1968)
	Astragalus erinaccus (L)	U.S.S.R.	Burenkov and Kuzina (1968)
	Astragalus microcephalus (L)	U.S.S.R.	Burenkov and Kuzina (1968)
Boron	*Eurotia ceratoides* (L)	U.S.S.R.	Buyalov and Shvyryayeva (1961)
	Limonium suffruticosum (L)	U.S.S.R.	Buyalov and Shvyryayeva (1961)
	Salsola nitraria (L)	U.S.S.R.	Buyalov and Shvyryayeva (1961)
Cobalt	*Crotalaria cobalticola* (U)	Katanga	Duvigneaud (1959)
	Silene cobalticola (U)	Katanga	Duvigneaud (1959)
Copper	*Acrocephalus robertii* (U)	Katanga	Duvigneaud (1959)
	Armeria maritima (L)	Wales	Henwood (1857)
	Astragalus declinatus (U)	U.S.S.R.	Malyuga et al. (1959)
	Becium homblei (U)	Zambia	Horizon (1959)
	Elsholtzia haichowensis (L)	China	Tsung-shan (1957)
	Elsholtzia patrini (L)	China	Chou-chin-han (1960)
	Eschscholtzia mexicana (L)	Arizona	Cannon (1960b)
	Gypsophila patrini (U)	U.S.S.R.	Nesvetaylova (1961)
	Merceya latifolia (U)	Sweden	Persson (1948)
	Merceya ligulata (U)	N. America	Shacklette (1967)
	Mielichhoferia macrocarpa (U)	Alaska	Shacklette (1967)
	Mielichhoferia mielichhoferi (U)	N. America	Shacklette (1967)
	Polycarpaea glabra (L)	Queensland	Nicolls et al. (1965)
	Polycarpaea spirostylis (L)	Queensland	Bailey (1899); Skertchly (1897)
	Polygonum posumbu (L)	China	Chou-chin-han (1960)
	Tephrosia sp.	Queensland	Nicolls et al. (1965)
	Viscaria alpina (U)	Norway	Vogt (1942a)
Iron	*Betula* sp. (L)	Germany	Lidgey (1897)
	Clusia rosea (L)	Venezuela	Buck (1951)
	Dacrydium caledonicum (L)	New Caledonia	Le Jolis (1860)
	Damnara ovata (L)	New Caledonia	Le Jolis (1860)
	Eutessa intermedia (L)	New Caledonia	Le Jolis (1860)
Lead	*Baptisia bracteata* (L)	Wisconsin	Cannon (1960b)
	Erianthus giganteus (L)	Tennessee	Cannon (1960b)
	Tephrosia polyzyga (L)	N. Australia	Cole et al. (1968)
Manganese	*Digitalis purpurea* (L)	U.S.S.R.	Uspensky (1915)
	Fucus vesiculolus (L)	U.S.S.R.	Malyuga (1947)
	Trapa natans (L)	U.S.S.R.	Malyuga (1947)
	Zostera nana (L)	U.S.S.R.	Malyuga (1947)
Molybdenum	*Astragalus declinatus* (L)	U.S.S.R.	Malyuga et al. (1959)
Nickel	*Alyssum bertolonii* (L)	Italy	Minguzzi and Vergnano (1948)
	Asplenium adulterium (L)	Norway	Vogt (1942a)
	Pulsatilla patens (L)	U.S.S.R.	Storozheva (1954)
Phosphorus	*Convolvulus althaeoides* (L)	Spain	Lidgey (1897)
Selenium	*Aster venusta* (U)	W. U.S.A.	Trelease and Beath (1949)
	Astragalus spp. (U)	W. U.S.A.	Trelease and Beath (1949)
	Neptunia amplexicaulis (L)	Queensland	McCray and Hurwood (1963)
	Oonopsis spp. (U)	W. U.S.A.	Trelease and Beath (1949)
	Stanleya spp. (U)	W. U.S.A.	Trelease and Beath (1949)

TABLE 4-1 (Continued)

Element	Species[a]	Locality	Reference
Selenium and Uranium	*Astragalus* (certain spp.)(U)	W. U.S.A.	Cannon (1957, 1960a)
Silver	*Eriogonum ovalifolium* (L)	Montana	Henwood (1857)
	Lonicera confusa (L)	Queensland	Bailey (1899)
Vanadium	*Astragalus bisulcatus* (U)	W. U.S.A.	Trelease and Beath (1949)
Zinc	*Gomphrena canescens* (L)	N. Australia	Cole et al. (1968)
	Matricaria americana (L)	Brazil	Dorn (1937)
	Philadelphus sp. (L)	U.S.A.	Cannon (1960b)
	Thlaspi calaminare (U)	Germany	Dorn (1937)
	Thlaspi cepaeaefolium (U)	S. Tyrol	Dorn (1937)
	Viola calaminaria (U)	Germany	Schwickerath (1931)
	Viola lutea (U)	Germany	Schwickerath (1931)

[a] Symbols in parenthesis after plant name signify whether local indicator (L) or universal indicator (U).

table has been brought up to date and includes species believed to have an indicator significance; one or two doubtful cases have been included for the sake of completeness.

4-1   FACT AND FOLKLORE ABOUT INDICATOR PLANTS

> Whosoever shall entertain high and vapourous imaginations, instead of a laborious and sober inquiry of truth, shall beget hopes and beliefs of strange and impossible shapes.
>
> Francis Bacon

Numerous references to indicator plants have been made during the past 150 years, but the compiler of a table of them is immediately confronted with the problem of what to include in the listing. It is not particularly important to adopt as a criterion the fact that a plant has or has not been used in prospecting work or that the results have or have not been successful, since if a plant is a true indicator of mineralization, the potential is always present. A far more serious problem is assessment of the credibility of an author's claims about a given species. Sometimes claims are lost in antiquity and are little more than folklore. Examples of folklore are the stories about plant indicators of diamonds and other precious stones (Spix and Martius, 1824).

Sometimes a plant can acquire the reputation of being an indicator or accumulator on very slender evidence. A classical example of this is the story of gold accumulation by the *Equisetum* (horsetail), a genus of some 25 species distributed throughout the world. Nemec et al. (1936) reported 610 ppm gold in the ash of *E. palustre* growing in soil containing only 0.2 ppm of this element and 63 ppm in a specimen of *E. arvense*. These values have been taken at their face value and have been quoted by many workers subsequently (Benedict, 1937; Lundberg, 1941; Thyssen, 1942; Babička, 1943; Malyuga, 1964). The original report has stimulated the use of *Equisetum* species in biogeochemical prospecting for gold (Warren and Delavault, 1950a). Subsequent workers have never found amounts of gold in horsetails even approaching the values reported by Nemec. Work by Warren and Delavault (1950a) and Razin and Roshkov (1963) gave values of less than 0.4 ppm for a total of 12 specimens.

Recently, Cannon et al. (1968) have carried out a very extensive investigation into the elemental composition of over 40 horsetails from the United States. In no case did the gold content exceed 0.5 ppm although there was ample evidence that horsetails are accumulators of zinc. The authors concluded that Nemec's original data for gold were incorrect and could have been due to an analytical technique which was likely to have serious interference from copper. The author agrees with the findings of Cannon and her co-workers and feels that the reputation of the horsetail as a gold indicator or accumulator is unwarranted and provides a classical example of the danger of accepting unsupported evidence in this type of work.

Dorn (1937) was one of the first to point out the full potential of the use of indicator plants. His paper mentions a number of species alleged to have been used successfully in mineral exploration; unfortunately it is entirely lacking in references. Six indicators of gold in Brazil are mentioned (all discovered by F. Freise). The same discoverer apparently found two other species said to be indicators of tin. The fact that numerous indicators of gold and tin have been found by one worker only and in one single country leads the reader to believe that the findings are in error, particularly as documentation is meager and there has been no confirmation of the "discovery."

Henwood (1857), in discussing a copper bog at Merioneth in Wales (see Chapter 17), wrote: "Persons conversant with the copper turbaries, consider the presence of metal in the soil indicated by the growth of Sea Pink (*Armeria maritima*), which appears to flourish there with remarkable luxuriance." The reputation of this plant as a copper indicator is based only on this obscure paper in which Henwood himself only quoted the opinion of the local people.

4-2   PLANT INDICATORS OF COPPER

There are frequent references to copper indicators in the literature (Henwood, 1857; Skertchly, 1897; Bailey, 1899; Vogt, 1942a; Lovering et al., 1950; Tsung-Shan, 1957; Duvigneaud, 1958, 1959; Horizon, 1959; Malyuga, 1959; Nesvetaylova, 1961; Cole et al., 1968). Usually, documentation is sufficiently dettailed to enable the reader to assess the value of the data except perhaps in the case of some of the nineteenth century work.

One of the most interesting cases of a copper indicator recorded in the early literature is that of *Polycarpaea spirostylis*, a member of the pink family found in Queensland. This plant has a reputation as an accumulator of copper (Skertchly, 1897; Bailey, 1899; Cole et al., 1968). Although further references in the literature are meager, extracts from Skertchly, quoted below, give a certain air of respectability to this plant as an indicator of copper:

> I venture to call it the "copper plant" and feel pretty sure it will be found a useful indication of copper lodes in North Queensland.... It was first discovered by Mr. Herschel Babbage ... in the year 1858 ... I found it quite

> plentiful at Montalbion, Muldiva, Calcifer and all over the copper region as far as I travelled, but always on or close to the copper deposits, or along water courses charged with copper in solution . . . my experience was that the plant could be detected more readily than the copper outcrop for they often perfectly swarmed and from their peculiar character were recognizable at a glance.

Further work in Australia has reaffirmed the likelihood that *P. spirostylis* is indeed a copper indicator.

Perhaps the most successful of all copper plants has been *Becium homblei*, discovered in Zambia in 1949. This plant belongs to the mint family, will not grow in soils containing less than 100 ppm copper, and will tolerate concentrations of over 5000 ppm (Horizon, 1959). Several important copper anomalies are said to have been discovered by use of this plant.

A very thorough geobotanical survey in Katanga (Duvigneaud, 1958, 1959) has produced evidence of several important copper indicators including *Acrocephalus robertii* and *A. katangaensis*. Other species such as *Silene cobalticola* and *Crotalaria cobalticola* have also a very high tolerance to cobalt, and it is claimed that they are indicators for this element.

Another well-documented copper plant is *Gypsophila patrini*, which is found in the Soviet Union. Nesvetaylova (1961) studied the distribution of this plant in soils of varying copper concentrations. The species apparently flourishes in soils averaging 1000 ppm base metals (copper, 300–1000 ppm), is totally absent at higher concentrations of these metals, and has an infrequent distribution for soils with a mean base-metal content of 40 ppm (copper, 30 ppm). At copper levels in the soil below 30 ppm, the species is again absent.

A striking example of a copper plant is *Eschscholtzia mexicana*, which acts as a local indicator in parts of Arizona. Figure 4-1 shows the marked predeliction of this species for mineralized ground.

4-3   PLANT INDICATORS OF SELENIUM

One of the most intensive and successful applications of the geobotanical method has been the use of selenium indicators on the Colorado Plateau (Cannon, 1952, 1957, 1960a, 1964). These plants have already been discussed in Chapter 3 under the heading of characteristic floras but further comment on them is merited.

Selenium indicators comprise *Aster venusta* and various species of *Astragalus*, *Oryzopsis*, and *Stanleya*. The presence of carnotite in seleniferous soils and rocks has an effect of increasing the availability of this element to plants as well as increasing the total amount of selenium in the substrate. This results in a marked tendency for species such as *Astragalus* to grow in these mineralized areas. Such species may be classified as primary indicators of selenium and secondary indicators of uranium.

A number of other plants were found in the vicinity of uranium ores in the Colorado Plateau, but their presence probably reflects pH changes or changes in the availability of major plant nutrients (Cannon, 1957). For example, large

FIGURE 4-1   *Eschscholtzia mexicana* (California poppy)
indicating copper mineralization bounded by a fault,
San Manuel, Arizona. Photograph supplied by courtesy
of H. L. Cannon and the U.S. Geological Survey.

amounts of calcium and phosphorus are available to plants near carnotite deposits
with a consequent increase in the frequency of occurrence of calciphilous species
such as *Mentzelia*, *Cryptantha*, and *Oenothera* and of phosphorus absorbers such
as *Eriogonum*, *Allium*, and *Calochortus*. In addition to the above types of indicator
there are a number of species tolerant of mineralized ground.

In her report on indicator plants, Cannon (1957) lists nine indicators known to
have been effective in locating uranium, 25 species known to favor mineralized
ground, and a further 16 indicators tolerant of mineralization. This work together
with that of other workers of the U.S. Geological Survey at Denver (Cannon and
Starrett, 1956; Kleinhampl and Koteff, 1960; Froelich and Kleinhampl, 1960) ably
demonstrates the potentiality of the geobotanical method when applied systematically by enthusiastic workers of different disciplines acting together in
cooperation.

4-4   PLANT INDICATORS OF ZINC

A number of plants have been classified as zinc indicators; they are
members of the zinc floras described in Chapter 3. These species are confined
mainly to Europe. Dorn (1937) in his well-known, though somewhat unselective,

listing of indicators has also included a number of poorly documented species from Brazil. The same author has included *Amorpha canescens* as a zinc indicator from the United States, although according to Cannon (1960b) it does not grow on mineralized ground.

Cole et al. (1968) have reported on the association of *Gomphrena canescens*, *Polycarpaea synandra* var. *gracilis*, and *Tephrosia polyzyga* with lead/zinc mineralization in the Bulman-Waimuna Springs area of Northern Territory, Australia. Although none of these species is officially classified as an indicator, the authors noted that *G. canescens* was confined to soils containing more than 5000 ppm zinc and that *T. polyzyga* was confined to substrates containing at least 1000 ppm zinc and 500 ppm lead.

4-5   OTHER INDICATOR PLANTS

A full discussion of all other known plant indicators is beyond the scope of this work. However, brief mention may be made of *Eriogonum ovalifolium* for prospecting for silver in Montana (Henwood, 1857), various boron indicators (Buyalov and Shvyryayeva, 1961), and *Astragalus declinatus* for molybdenum (Malyuga, 1959). A surprising feature of the literature is that so many of the reliable species were known many decades ago and that the literature of the past five years contains few additional data on newly discovered indicator plants. A rather novel approach has been that of Chiu and Wang (1964) who apparently used the color of certain soil bacteria as an indication of copper levels in the substrate.

Plant indicators of bitumen have also been reported (Yaroshenko, 1932; Popov, 1949). The species in all cases being confined to bituminous volcanic mud ejecta in Eastern Siberia.

4-6   BRYOPHYTES AS INDICATORS OF MINERALIZATION

Bryophytes (mosses and liverworts) and lichens have an extraordinary capability of absorbing trace elements from the substrate upon which they are growing and will often tolerate adverse ecological conditions to a much greater degree than vascular plants. The accumulation of trace elements by bryophytes is usually much greater than for other plant groups as has been shown by extensive studies on mosses and lichens in Finland (Lounamaa, 1965) and in North America (Shacklette, 1961; Leroy and Koksoy, 1962; Shacklette, 1965a, 1965b, 1967).

Specialized types of bryophytes are the so-called *copper mosses* which are reputed to grow only on substrates high in copper. The predilection of these bryophytes for copper has been known ever since their discovery in the early nineteenth century, though a fuller report on them did not appear until 100 years later (Morton and Gams, 1925).

Copper mosses belong mainly to two genera: *Mielichhoferia* and *Merceya*. Mårtensson (1956) and Persson (1956) reported on copper levels in substrates of *Mielichhoferia elongata* growing in Scandinavia and gave values ranging between 20–450 ppm with most values in the higher part of the range. It has been reported that specimens of *Mielichhoferia macrocarpa* and *Mielichhoferia mielichhoferi* in North America (Schofield, 1959), grow on substrates rich in pyrite. The genus

Merceya has been described by Noguchi (1956) and Persson (1948) and it is believed that all of its six species are probably cuprophile.

There has been some controversy for many years as to whether in fact the response of copper mosses is directly to this metal itself or whether it is a function of the pH or sulfide content of the substrate (Schatz, 1955). In a recent paper Hartman (1969) has ascribed all these variables to the ecology of *mielichhoferia* in Colorado. The presence of a low pH over copper deposits does not, however, always seem to be a significant feature because Persson (1956) has reported a high copper concentration (320 ppm) in the substrate of *Merceya latifolia* together with a high pH (7.6).

Recently, Shacklette (1967) has carried out extensive analyses of mosses and their substrates collected from Alaska and of substrates of *Mielichhoferia mielichhoferi* growing in Michigan. He also obtained data for substrates of *M. elongata* from Sweden and France. His findings strongly suggest copper as the controlling element and tend to disprove the role of sulfur in affecting the distribution of these bryophytes.

Lichens have also been used in prospecting for minerals. Viktorov (1956) has reported that *Collema minor* and *Lecidea decipiens* are indicators of gypsum in desert areas of the Soviet Union.

It is clear that it will never be a practical proposition to use lichens or copper mosses directly as prospecting guides because, not only are they extremely rare, but they are also difficult to identify for anyone but a skilled bryologist. Nevertheless, Persson (1948, 1956) has suggested a novel method of using these plants to detect mineralization. The procedure involves examining specimens of copper mosses at present held in many university and government herbaria, noting the locality from which they were obtained, and then reexamining the location by other prospecting methods. Cannon (1960b) has stated that three copper deposits have been found by this method.

4-7   COMMON CHARACTERISTICS OF INDICATOR PLANTS

Few new indicator plants have been discovered in recent years. This is probably partly due to a trend of greater insistence on confirmatory data before assigning indicator characteristics to a plant and perhaps partly due to the fact that the most obvious indicators have already been discovered so that basic ground rules must be established in the search for new indicators.

In surveying the literature, the reader is immediately impressed by a fairly consistent pattern in most indicator plants. Excluding bryophytes, the typical indicator will be a herb rather than tree or shrub, will probably be a member of the Labiaceae (mint) or Caryophyllaceae (pink) families, and will most likely be more effective for copper than for other elements. A universal characteristic of indicators is that they will have a very high elemental content in their ash. Table 4-2 lists maximum values, for the content of various elements in indicator plants. In nearly all cases values are at least an order of magnitude greater than normal values for unmineralized soils. This is not unexpected in view of the fact that indicator plants are often

TABLE 4-2   Elemental Accumulations in the Ash of Indicator Plants

Element	Species	Normal content in vascular plants[a] (ppm)	Maximum content in indicators (ppm)	Reference
Cobalt	*Crotalaria cobalticola*	9	18,000	Duvigneaud (1959)
Copper	*Becium homblei*	183	2,500	Horizon (1959)
	Gypsophila patrini	183	500	Nesvetaylova (1961)
Manganese	*Fucus vesiculolus*	4,815	90,000	Malyuga (1959)
Nickel	*Alyssum bertolonii*	65	100,000	Minguzzi and Vergnano (1948)
Selenium	*Astragalus pattersoni*	1	46,000	Cannon (1960a)
Zinc	*Thlaspi calaminare*	1,400	10,000	Dorn (1937)

[a] Cannon, 1960b.

those with the greatest tolerance to mineralized ground and the least tolerance to competition from other species. Investigations on a serpentine flora in New Zealand (Lyon et al., 1968) have clearly shown that the more tolerant a plant is to its environment, the higher will be its uptake of elements for which tolerance is required.

A new approach to the discovery of new indicator plants may well be the chemical analysis of all species growing near a mineralized area and subsequent selection for further testing of those species with the highest concentrations of the element sought.

4-8   EVALUATION OF SUCCESS IN PROSPECTING WITH INDICATOR PLANTS

An important part of geobotanical prospecting is the final evaluation of the success achieved, and there are various ways of assessing this. The apparent degree of success will unfortunately depend to some extent on the procedure used for the evaluation, and it is very difficult to avoid unconscious bias when selecting a suitable criterion for this purpose.

In work carried out in the Yellow Cat area, Utah, Cannon (1964) was able to compare data on indicator plants with the results of drillings carried out over the whole of the 6-sq mi area. Her method of assessment of success is presented as an example of a suitable procedure that can be followed.

Table 4-3 shows the distribution of indicator plants (*Astragalus pattersoni* and *A. preussi*) in the vicinity of drill holes driven into favorable and nonfavorable ground. Of the holes drilled in ground supporting indicator plants, 97% were later proved to have been driven in ground which was geologically favorable or semi-favorable. Indicator plants were clearly extremely useful in locating favorable ground though not necessarily mineralization within it.

The next stage in Cannon's work was to establish the effectiveness of the indicators in the geologically favorable ground. The drill holes were classified as

TABLE 4-3 Association of Indicator Plants with Geologically Favorable or Nonfavorable Ground in the Search for Uranium

Presence or absence of indicator plants	Number of drill holes		
	Semifavorable and favorable ground	Unfavorable ground	Total number
Present	311 (206)[a]	12 (117)	323
Absent	497 (602)	448 (343)	945

[a] Figures in parenthesis indicate the distribution that would be expected if there were no relationship between plants and ore bodies.

SOURCE: Cannon (1964).

TABLE 4-4 Association of Indicator Plants with Mineralized Ground in the Search for Uranium

	Number of drill holes			
	Ore-bearing ground	Mineralized ground	Unmineralized ground	Total
Indicator plants present	51 (17)[a]	99 (46)	113 (200)	263
Indicator plants absent	30 (64)	117 (170)	838 (751)	985
Total holes drilled	81	216	951	1248
Percentage of holes indicated as botanically favorable	63	46	12	

[a] Figures in parentheses indicate the distribution that would be expected if there were no relationship between plants and ore bodies.

SOURCE: Cannon (1964).

TABLE 4-5 Comparison of Indicator Plant Distribution with Drilling Results at Various Depths

Depth drilled (feet)	Number of holes[a]	Percentage of mineralized holes in which plants present	Percentage of ore holes in which plants present	Percentage of mineralized + ore holes in which plants present
0–9	16	54	60	54
10–20	35	78	33	70
21–32	32	75	100	81
33–49	32	41	60	46
50–67	33	38	56	42
68–99	34	44	44	43
100–115	36	28	78	42
116–150	35	29	81	45
151–169	31	36	41	38
170+	12	22	0	16
Totals	296	46	62	50

[a] Of 971 barren holes, only 14% had indicator plants growing near them.

SOURCE: Cannon (1964).

ore-bearing (more than 0.1% U_3O_8), *mineralized* (more than 0.02% U_3O_8), and *non-mineralized*. The data are shown in Table 4-4. It may be seen that indicator plants were found near 63% of the ore holes, 46% of the mineralized holes, and only 12% of the nonmineralized drillings.

The above evaluation had so far not considered the depth at which mineralization was found. Clearly there must be some limit to the effective depth of the method in any particular area. Table 4-5 shows data for the distribution of indicator plants in relation to results of drilling. The table shows that indicator plants favored ore holes even when the ore body was as deep as 170 ft. The greatest effectiveness was apparently for depths between 21–32 ft.

Many other methods have been used to evaluate success in geobotanical prospecting with indicator plants. In the Soviet Union use has been made of the so-called *reliability index* and *validity index* (Viktorov et al., 1962) which involve not only a study of the distribution of indicator plants but also their uptake of the element which is sought.

SELECTED REFERENCES
General: Cannon (1960b); Malyuga (1964); Nesvetaylova (1961)
Bryophytes as copper indicators: Persson (1948); Shacklette (1967)
Copper and cobalt indicators: Duvigneaud (1958, 1959)
Selenium indicators: Cannon (1957, 1960a)
Evaluation of effectiveness of indicators: Cannon (1964)

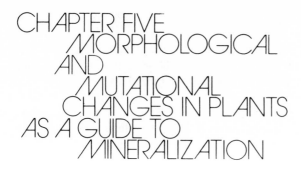

CHAPTER FIVE
MORPHOLOGICAL
AND
MUTATIONAL
CHANGES IN PLANTS
AS A GUIDE TO
MINERALIZATION

Changes in the morphology of plants and evidence of disease are useful aids in geobotanical prospecting and have been used as field guides since the eighteenth century. Early workers had to rely on obvious changes such as dwarfism or variation in color. But with the increasing sophistication of modern science and greater knowledge of plant physiology, many other visual indications of mineralization have been noted and can be used in prospecting.

Morphological changes in plants under the influence of mineralization are very varied and include such factors as *dwarfism, gigantism, mottling or chlorosis of leaves, abnormally shaped fruits, changes of color in the flowers, disturbances in the rhythm of the flowering period, changes in growth form*, and a large number of other indications. Considerably more skill is required for recognizing morphological changes than is required for studying the distribution of indicator plants (see Chapter 4) unless the changes are very obvious. Usually a trained botanist or plant physiologist should be employed for this type of work. Some common morphological and mutational changes in plants are listed in Table 5-1 and will be considered in more detail below.

5-1   **TOXICITY IN PLANTS**
Most morphological and mutational effects in plants may be said to be due to the toxic effects of minerals in the substrate. There is extensive literature on the subject of toxicity of elements to plants and it is possible to make general conclusions concerning the effects of certain metals in plants and to establish basic reasons for the effects which are produced.

Bowen (1966) has suggested that elements may be divided into three classes according to their toxicities.

1. *Very toxic:* toxicity symptoms appear at concentrations less than 1 ppm in the substrate. Such elements include beryllium, copper, mercury, silver, and tin.

TABLE 5-1   Morphological and Mutational Changes in Plants Affected by Mineralization

Element or mineral	Effect	References
Aluminum	Shortening of roots; leaf scorch	Wallace (1951)
Bitumen	Gigantism, early or second flowering	Vostokova et al. (1961)
Boron	Stunting; prostrate forms; deformation; browning and blotching of leaves	Buyalov and Shvyryayeva (1961); Wallace and Bear (1949); Williams and Vlamis (1957)
Chromium	Chlorosis of leaves	Hewitt (1953)
Cobalt	Increase of chlorophyll in some species and chlorosis in others	Duvigneaud (1959); Hewitt (1953)
Copper	Chlorosis of leaves and dwarfism	Duvigneaud (1959); Hewitt (1953)
Iron	Darkening of leaves	Burghardt (1956)
Manganese	Chlorosis of leaves with white blotches	Löhnis (1950, 1951)
Molybdenum	Formation of abnormally colored shoots	Warington (1937)
Nickel	Chlorosis and necrosis of leaves	Hewitt (1953)
Serpentine	Dwarfism; color changes of flowers	Krause (1958)
Uranium and radioactivity	Variation in flower color; presence of abnormal fruits; increase in chromosomes of nucleus; stimulation	Cannon (1960b, 1964); Shacklette (1962a, 1964)
Zinc	Chlorosis of leaves; symptoms of manganese deficiency	Hewitt (1953)

2. *Moderately toxic:* toxicity symptoms appear at concentrations between 1–100 ppm in the substrate. Examples of this are the transition elements and most of the elements of Groups III, IV, V, and VI of the periodic table.

3. *Scarcely toxic:* toxicity symptoms rarely appear at concentrations normally encountered in the substrate. Examples of such elements are the halogens, nitrogen, phosphorus, sulfur, titanium, the alkali metals, and the alkaline earths.

The above classification applies to vascular plants rather than bryophytes and the data are based on concentrations of the elements in nutrient solutions where the whole concentration is assumed to be available to the plant. Where plants are found under natural conditions, substantially higher concentrations of these elements can be tolerated in the soil, provided that availability is low. With regard to the very toxic element beryllium, it is worthy of record that amounts of up to 27 ppm of this element have been reported in the ash of *Coprosma australis* growing in a uraniferous area in New Zealand (Cohen et al., 1967).

The most common mechanisms of toxic action in plants are as follows:

POISONING OF ENZYMES   The more electronegative metals such as copper, mercury, and silver have a great affinity for sulfhydryl groups, which are reactive sites for many enzymes. The enzyme is therefore unable to function and toxicity results. As might be expected there is an approximate relationship between the toxicity of an element to plants and its electronegativity (Somers, 1960). For divalent metals the order of electronegativity decreases in the sequence: mercury, copper, tin, lead, nickel, cobalt, cadmium, iron, zinc, manganese, magnesium, calcium, strontium, and barium so that mercury is the most toxic of the series and barium the least toxic. It must be emphasized that this rule is only very approximate and differs with species.

REPLACEMENT OF ESSENTIAL NUTRIENTS   Arsenate and chlorate can occupy sites normally involving phosphate and nitrate respectively (Åberg, 1948). Phosphorus and nitrate deficiency symptoms can therefore appear in the plant.

PRECIPITATION OF ESSENTIAL NUTRIENTS   Elements such as aluminum, beryllium, and titanium readily precipitate phosphate, rendering it unavailable to the plant. In this way phosphorus deficiency as above can result.

CATALYTIC DECOMPOSITION OF ESSENTIAL NUTRIENTS AND METABOLITES   Elements such as lanthanum have a strong catalytic action on the decomposition of metabolites such as ATP (Bowen, 1966).

COMBINATION WITH THE CELL MEMBRANE AND REDUCTION OF ITS PERMEABILITY   Elements such as copper, gold, lead, and mercury are able to reduce the permeability of the cell membrane (Passow et al., 1961) and prevent the free passage of potassium, sodium, or organic molecules.

REPLACEMENT OF STRUCTURALLY IMPORTANT ELEMENTS IN THE CELL   Some elements replace others in the cell but have an inert physiological role. Examples of this are the replacement of sodium by lithium and of chlorine by bromine (Pirschle, 1930, 1932).

It is obvious that a further discussion of toxicity is beyond the scope of this work, but it is hoped that the above discussion will be of assistance in providing workers in this field with a few basic ideas concerning the relationship of the elements in plants and soils. For a further discussion of toxicity of the elements to plants, the reader is referred to Hewitt and Nicholas (1963) and Bowen (1966).

An examination of Table 5-1 will show that there are so many different morphological changes produced in plants by mineralization in the substrate that not even an expert could be expected to realize the full significance of each variation in vegetation. Basically however, most changes may be classified under the headings of: abnormality of form, chlorosis of leaves, color changes in flowers, dwarfism, gigantism, and rhythmic changes in flowering periods. Each will be considered in turn.

5-2   **ABNORMALITY OF FORM**

Abnormality of form, excluding dwarfism or gigantism, is usually a sign of the presence of either boron or of radioactive minerals. The work of Buyalov and Shvyryayeva (1961) on the affect of boron on vegetation in the Soviet Union serves as a model for the successful use of morphological changes in plants in the search for minerals. These authors noted various variations in the vegetation depending on the boron concentration in the soil. At intermediate boron concentrations, individual plants such as *Artemisia lercheana* and *Kochia prostrata* were larger and had fresher vegetation than plants growing over soils with normal boron contents. This effect was particularly striking in summer when so much of the other vegetation was desiccated. At high boron levels in the soil, toxicity symptoms began to appear in the vegetation. This took the form of dwarfism and deformation of species such as *Eurotia ceratoides*, *Anabasis salsa*, and *Salicorna herbacea*. In some cases a prostrate form became apparent and there was increased branching of the stems. At very high concentrations of boron the only surviving species were the indicator plants *Limonium suffruticosum* and *Salsola nitraria* (see Chapter 4). Similar findings were reported by Kantor (1959) who studied deep-rooted species in the Soviet Union and found that positive responses could be obtained for boron for depths of up to 90 ft.

Unusual and unpredictable changes of form are produced by radioactivity. The first result of mild doses of radioactivity is a stimulatory effect on the vegetation. After the nuclear explosion at Hiroshima, exceptional yields of various crops were obtained in the following season (Cannon, 1964); there were also several cases of grotesque mutations in the vegetation.

Fortunately, however, natural radiation is never as high as that encountered at Hiroshima in 1945 and levels normally encountered are therefore seldom sufficient to produce an obvious stimulatory effect on vegetation. There is, however, ample evidence that even fairly low levels of radiation can produce morphological changes in plants over a prolonged period. Shacklette (1962a) has described variations in the fruit of the bog bilberry (*Vaccinium uliginosum*) growing in a radioactive area at Great Bear Lake in Canada. Six main variant fruit forms were found and are illustrated in Figure 5-1. The colony of unusual variants was situated directly above a large deposit of pitchblende which outcropped on each side of it, and it is likely that Shacklette's hypothesis that mutations were due to radioactivity is correct.

In experiments with plants grown artificially in carnotite, Cannon (1960a) reported the usual stimulatory effect in many species. Unusual plant forms included enlargement of the basal root stem of *Grindelia* in which the basal rosette of leaves was raised 12 in. from the ground. *Stanleya pinnata* growing in the same plots produced stalks of imperfect flowers having no petals or stamens and carrying greatly enlarged green sepals.

5-3   CHLOROSIS OF LEAVES

One of the commonest field guides to mineralization is the presence of chlorosis in the vegetation. This characteristic "yellowing" is nearly always an indication of iron deficiency in the plant. It is caused very rarely by low levels in the substrate; rather, it is a result of the antagonistic effect of other metals towards uptake of iron by the plant (see Figure 3-4). Chlorosis is also a symptom of manganese deficiency, although in this case the effect is one of chlorotic patches or streaks rather than an overall effect in the leaves (Stiles, 1958).

As can be seen from Table 5-1, chlorosis is an indication of excessive amounts

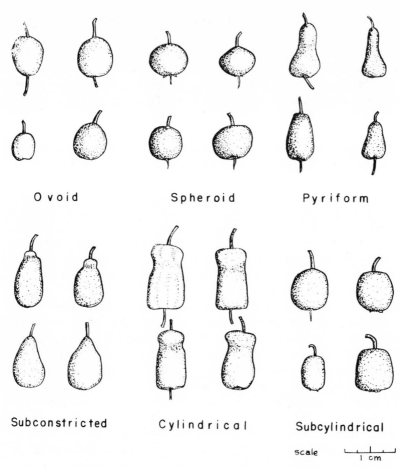

O void   Spheroid   Pyriform

Subconstricted   Cylindrical   Subcylindrical

scale   ⊢——⊢——⊣
              1 cm

FIGURE 5-1 Fruit variation in *Vaccinium uliginosum* L. The labeled groups represent the range of fruit shapes borne on a single clone. The upper right drawing in each group represents the typical fruit shape of the clone, and the other three drawings represent variants. All drawings were made from fresh, mature fruits. From Shacklette (1962a). By courtesy of the Canadian Field Naturalist.

of chromium, cobalt, copper, manganese, nickel, or zinc in the soil as all these elements are antagonistic to iron. The observation of chlorosis in leaves is therefore of little use to the prospector in deciding which element is present in excessive amounts in the soil but does at least indicate that some type of mineralization may be present.

5-4   COLOR CHANGES IN FLOWERS
        Perhaps the most obvious of all plant mutations is that of changes in the color of the flowers. Provided that a field trip can be carried out during the flowering period, very effective results can be achieved in a short space of time.

Color changes in flowers are usually the result of either radioactivity or of the presence of certain elements in the soils. Shacklette (1964) has observed extensive flower variation of *Epilobium angustifolium* growing in Canada over uranium deposits adjacent to these described in Section 5-2. These variations are shown in Table 5-2, which compares data for eight specimens of radioactive plants and two normal individuals. The table gives some indication of the factors that can be recorded by a trained geobotanist, but it is obvious that work of this nature is really the domain of the expert rather than the amateur.

In looking at color variations, due attention must be paid to the normal variability of this factor in background areas since geobotanical interpretation must depend on a greater-than-average frequency of occurrence in the area under study. If precise values cannot be ascribed to the factor being considered, it is usually sufficient to assign terms such as rare, infrequent, common, and abundant to its frequency of occurrence.

TABLE 5-2 Mutational Changes in the Color of Flowers of *Epilobium angustifolium* (from Port Radium, Canada) Under the Influence of Radioactivity (Nos. 1 to 8) Compared with Normal Specimens from Alaska (Nos. 9 and 10)

Clone no.	Petals	Sepals	Filament	Anther	Pollen	Capsule
1	Pale rose pink	Cerise	Pure white	Dusky pink	Pale blue	Red above, paler beneath
2	Magenta, paler than typical	Reddish-purple	Pure white	Yellow	Very pale blue	Paler than usual
3	Very pale rose pink	Clear rose red	White, pink at base	Purplish	Greenish-blue	Dusky pink
4	Very pale pink	Pale to clear rose pink	White, rose at base	Purplish	Bluish-green	Spreading; shorter than normal
5	Intense magenta; deep fuchsia	Darker than petals, purplish	Pure white	Purple	Very pale, slightly bluish	Erect; pale pink
6	Most intense magenta of entire population	Very slightly darker than petals	Pure white	Reddish	Pale blue	Erect; medium length
7	Pale magenta	Rosy-purple	Pure white	Yellow	Blue	Erect
8	Pale magenta	Rosy-purple	Pure white	Yellow	Blue	Erect
9 (normal)	Uniform light pink	Clear red	White, to pale pink at base	Pale pink	Yellowish-green	Ascending; red above, pale green beneath
10 (normal)	Magenta	Dark magenta	Very pale pink, turning white	Greenish	Light blue-green	Erect; reddish above, greenish beneath

SOURCE: Shacklette (1964).

Metal ions as well as radioactivity can affect the color of flowers. The gardener's trick of adding iron or aluminum to red hydrangeas to turn them blue is of course well known. The theory behind such color changes is interesting and may have bearing on mineral exploration.

The majority of flower colors are produced by a surprisingly small number of pigments. Apart from *carotenoids*, which are important in yellow and orange flowers, it is mainly the *anthocyanins* that are responsible for the color range from orange to deep blue. Bayer et al. (1966) have suggested that in the absence of certain metals, the anthocyanins form red *oxonium salts* which become blue when they are complexed with excessive amounts of iron, aluminum, or other elements. This is precisely the mechanism involved in the color change of the hydrangea referred to above.

Besides iron and aluminum, other elements such as chromium, tin, titanium, and uranium can form stable complexes with anthocyanins. It is therefore possible that excessive amounts of some of these metals could produce a blue tint in flowers that are normally red or pink, and this could be a useful field guide in prospecting. The author has observed a blue-red form of the manuka, *Leptospermum scoparium*

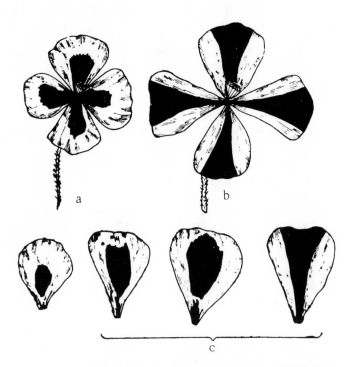

FIGURE 5-2   Change in the color of the petals (shading) of the flower of the poppy *Papaver commutatum* F. et M. under the influence of copper-molybdenum mineralization. (a) Normal flower; (b) modified flower; (c) degree of mutability of petals of the corolla. From Malyuga (1964). By courtesy of the Plenum Publishing Corp.

(normal color white or pale pink), growing in soil containing 6% chromium in an ultramafic area in New Zealand.

There are several other references in the literature to color changes produced in flowers by mineralization in the substrate. Examples are: color variations in the *Eschscholtzia* (Bazilevskaya and Sibireva (1950) and variations in flowers of the poppy, *Papaver commutatum* (Malyuga et al., 1959). This latter mutation, said to be due to copper or molybdenum, is illustrated in Figure 5-2.

5-5   DWARFISM

Stunting of growth, or dwarfism, is one of the commonest toxicity manifestations in vegetation and is moreover very difficult to assign to any particular cause since it is likely to be due to deficiency or excess of any one of a number of different elements. However, the presence of ultrabasic rocks in the substrate is invariably indicated by dwarfism of serpentine floras. One of the most marked examples of dwarfism known to the author is that of the New Zealand shrub *Pittosporum rigidum*, which under normal conditions grows to 15 ft in height but in serpentine areas is a cushion plant a few centimeters high (Bell et al., 1911).

5-6   GIGANTISM

Gigantism is considerably less common than dwarfism and is usually associated with the presence of bitumen or boron. Care must be taken not to confuse gigantism with the stimulatory effects produced from an abundance of plant nutrients or even from the presence of radioactivity.

There are frequent references in the literature to the beneficial effects of petroleum products on vegetation. These effects are produced partly by conditioning mechanisms in which the structure of the soil is improved so that the temperature and moisture capacity are increased. Other effects include an increase in the availability of phosphate and in the intensity of nitrification (Remezov, 1938). In certain cases, however, petroleum products can cause abnormalities in vegetation resulting typically in gigantism. One of the first records of this type of effect was by Shchapova (1938), who found forms of *Zostera nana* five times their normal size growing in a bituminous area on the shores of the Caspian Sea.

A very thorough investigation of the effects of petroleum on vegetation was later carried out by Vostokova et al. (1961) in West Kazakhstan and on the coast of the Caspian Sea. These workers observed gigantism in 29 species. Figure 5-3 shows typical data for *Atriplex cana* for a bituminous soil and a background area as control. It is clear that the two areas can be completely characterized from the height and diameter data.

The above species were those showing uniform gigantism in all dimensions. Other plants showed increased height with no lateral increase in dimensions and others showed the opposite effect. In both cases the species could be used to characterize soils with a high bitumen content.

The above work has demonstrated that the geobotanical method can be used even in the search for petroleum and is not restricted purely to inorganic constituents of the soil and bedrock.

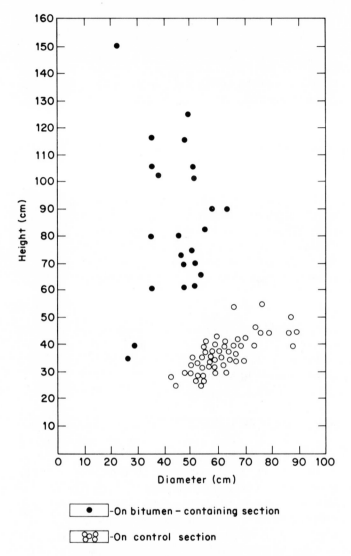

FIGURE 5-3 Graphical representation of size ratios of
Atriplex cana growing over bituminous and nonbitumi-
nous areas. From Vostokova et al. (1961). By courtesy of
the American Geological Institute.

5-7 RHYTHMIC DISTURBANCES IN PLANTS

One effect of mineralization on plants is a disturbance of the rhythm
of flowering and foliation. Sometimes flowering occurs earlier or later; in other
cases there is a second flowering, usually in the fall. Examples of this have been
reported by Buyalov and Shvyryayeva (1961) for plants growing over boron deposits in the Soviet Union. Vostokova et al. (1961) have reported a similar effect

for vegetation growing in soil containing bitumen. Second blooming was noted for 12 species although not every individual would show this effect. Late blooming was also observed for *Stanleya pinnata* exposed to radioactivity (Cannon, 1960a).

The rhythmic disturbances referred to above all have one feature in common. They were all caused by minerals whose main effect was that of stimulation rather than depression. It is probable, therefore, that in field work abnormal flowering times may be linked to the presence of such stimulants as bitumen, boron, or radioactivity.

5-8   CONCLUSIONS

Certain general conclusions may now be made concerning morphological changes in plants as a guide to mineralization. It is clear that to identify many of these changes, the services of a skilled botanist or ecologist will be required. This is particularly true for the recognition of plant abnormalities and for the proper understanding of the true significance of chlorotic effects in leaves. Other variations such as dwarfism, gigantism, and color changes in flowers may be readily observed by less skilled personnel.

It is concluded that most chlorotic symptoms will be associated with iron deficiency occasioned more likely by antagonistic effects from other elements rather than as a result of iron deficiency in the substrate.

Early and second flowering are probably indications of stimulants such as excess of certain nutrients, e.g., boron, nitrogen, phosphorus, or potassium, or may be due to the presence of bitumen or radioactivity. Lateness in flowering may, however, be a toxicity symptom linked with the vigor of the plant.

When used in conjunction with the study of indicator plants and characteristic floras, the study of morphological changes should be an effective aid in the application of geobotany to the search for minerals.

SELECTED REFERENCES
General: Cannon (1960b); Malyuga (1964)
Chlorosis: Hewitt (1953)
Dwarfism: Krause (1958)
Effect of radioactivity: Cannon (1964); Shacklette (1962a, 1964)
Gigantism: Vostokova et al. (1961).

CHAPTER SIX
AERIAL
GEOBOTANICAL
SURVEYS

In recent years the development of more sophisticated aircraft, new techniques in remote sensing, and the birth of the space age have introduced new and exciting dimensions to indicator geobotany. Greater pressure on natural resources caused by an expanding population and rising standards of living throughout the world has led to geobotanists taking to the air in ever-increasing numbers in order to carry out an inventory of the vegetation cover and to assess its significance in the search for water, petroleum, and minerals.

The development of artificial satellites during the past decade has opened up limitless possibilities in aerial mapping. Technological progress has also provided greatly improved techniques for remote sensing of natural resources by use of detectors sensitive to almost the whole of the electromagnetic spectrum from gamma radiation at the one extreme to radar at the other (Colwell, 1968). Figure 6-1 shows the wavelengths commonly employed in various types of remote sensing and shows their relative absorption by the atmosphere, which is of course the limiting factor in their use.

Table 6-1 lists six techniques of remote sensing arranged in order of increasing wavelength of the radiations used. Each of these methods will be considered in turn and its potential in indicator geobotany will be evaluated.

6-1   AERIAL GAMMA SCINTILLOMETRY AND GAMMA SPECTROMETRY

Gamma scintillometry and gamma spectrometry do not at first sight appear to involve vegetation to any great extent and their inclusion in a work on geobotany and biogeochemistry would not therefore seem to be justified. There is, however, evidence that vegetation itself can contribute to, or interfere with, the gamma radioactivity of heavily forested areas (Balabanov and Kovalevsky, 1963). A further factor that should be mentioned is that the presence of thick vegetation may itself render an aerial gamma survey necessary if access on the ground is restricted.

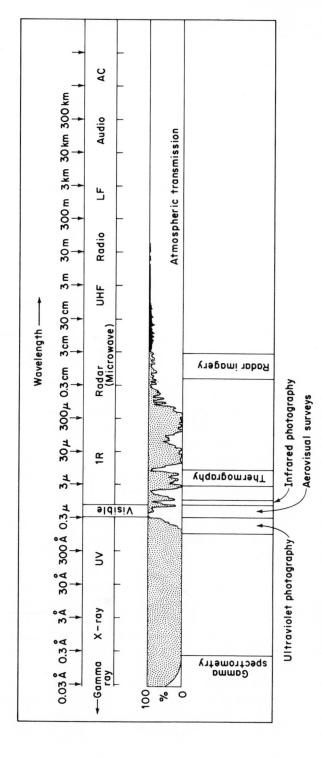

FIGURE 6-1 Schematic representation of the electromagnetic spectrum and its significance in remote sensing. After Colwell et al. (1963). By courtesy of the American Society of Photogrammetry.

TABLE 6-1   Remote Sensing at Various Bands of the Electromagnetic Spectrum

Technique	Wavelength	References
Gamma spectrometry and gamma scintillometry	1 Å	Balabanov and Kovalevsky (1963); Pitkin et al. (1964); Stubbs (1968)
Photography in visible and ultraviolet range	1800–8000 Å	Colwell (1964); Ray (1960)
Aerovisual surveys	4000–8000 Å	Kasynova (1961); Viktorov et al. (1964)
Infrared photography	7000–9000 Å	Colwell (1964, 1967, 1968)
Thermography	30,000–150,000 Å	Colwell (1967, 1968)
Radar imagery	1–3 cm	Colwell (1967, 1968)

The detector used in both radiometric methods is usually a thallium-activated sodium iodide crystal which receives the gamma radiation and converts it into electrical energy via a photomultiplier and amplification system.

Aerial gamma scintillometry (Pitkin et al., 1964) has been used successfully for many years in various parts of the world. In a typical project carried out recently in New Zealand, an extensive and very mountainous area was systematically mapped from a scintillometer carried aboard a helicopter. The area was surveyed by flying above the ground contours at 100-ft height intervals. Constant monitoring of activity was carried out with a recorder, and the values were plotted on a map as required. An alarm bell fitted to the system gave a signal when activities exceeded a certain predetermined level so that more frequent readings could be made of the area concerned. Several of the "anomalies" detected from the air proved to be false and due largely to the *mass effect* which can be a problem over very mountainous country (Whitehead and Brooks, 1969a). A scintillometer carried over the floor of a narrow valley will receive more radiation than one traveling over a ridge even if the amounts of radiation from below are the same, due to the lateral contributions from the valley walls.

A disadvantage of gamma scintillometry is that it is nonselective and will not distinguish between radioisotopes of potassium, thorium, or uranium, all of which could contribute to the activity recorded by the detector. This problem has been overcome by the use of airborne gamma spectrometry (Stubbs, 1968), which is able to differentiate between the above radioisotopes, all of which have different energies. The instrument divides the energy into the total gamma counts between 0.1 and 2.9 MeV (potassium = 1.45 MeV; uranium = 1.76 MeV; thorium = 2.62 MeV) and the partial gamma activities in the ranges 1.3–1.6 MeV, 1.6–2.0 MeV, and 2.0–2.9 MeV. Tracings for each energy range are shown on a recorder. Airborne gamma spectrometry is now becoming commercially available and has already been used successfully in the field. Figure 6-2 shows the results of an airborne survey carried out in Texas. A further advantage of gamma spectrometry is that it may be used to distinguish between different rock types by means of their varying uranium:thorium ratios.

There are conflicting reports on the effect of vegetation on gamma scintillometry. Balabanov and Kovalevsky (1963) claim that in wooded areas the trees may shield ground anomalies and can themselves provide false anomalies due to their uptake of radium-226. Although this last problem could probably be circum-

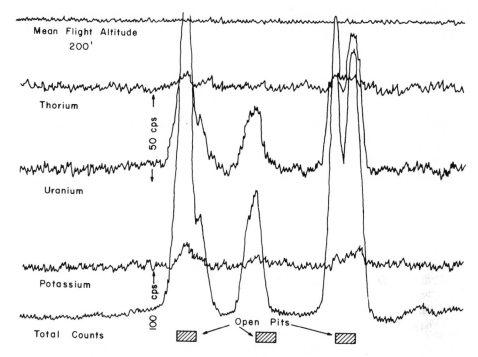

FIGURE 6-2   Four-channel airborne gamma spectrometer results over open-pit uranium deposits, South Texas. From Saum and Link (1968). By courtesy of the Colorado School of Mines.

vented by using gamma spectrometry, the question of gamma radiation absorbed by vegetation still remains. Confirmation of the shielding of gamma radiation by forests was obtained by Matveev (1961) who placed identical sources of cobalt-60 in forested areas and in open ground and then measured their activity by airborne scintillometry.

6-2   AERIAL PHOTOGRAPHY IN THE ULTRAVIOLET AND VISIBLE SPECTRAL RANGE

Aerial photography in the ultraviolet and visible parts of the electromagnetic spectrum has been carried out for over a hundred years. The earliest work of this kind was carried out from hot-air balloons, kites, and even from homing pigeons with miniature cameras attached to their legs. During both world wars, the technique was used extensively in military reconnaissance, but it was not until after World War II that the procedure began to be applied to any great extent in vegetation surveys. Extensive work in this field has now been carried out in the United States and in the Soviet Union. Most developed countries now have a complete range of aerial photographs for the whole of their territory.

The interpretation of aerial photographs in terms of geology is known as *photogeology* (Ray, 1960) and is finding increasing use among geologists and geomorphologists. In the Soviet Union, a clear distinction is made between the use of aerial photographs for indicator geobotany (Viktorov et al., 1964) and aerovisual surveys of an area with the aid of photographs carried in the aircraft (Kasynova, 1961).

The association between the geology and the vegetation of an area, as discussed in Chapter 3, has been utilized in the preparation of geobotanical maps from aerial photographs. The Soviet Union appears to be the most advanced in this field and has excellent geobotanical maps for a large part of its territory (Shvyrya-yeva and Mikhailova, 1965). Some geobotanical mapping has also been carried out in the United States and in Australia. The usefulness of aerial photographs (particularly in color) for this type of mapping can be gauged from the following information from Viktorov et al. (1964).

The color of vegetation in arid areas in the summer depends to a great extent on the availability of ground water. Color photographs of plant communities taken during the summer and winter months can be compared. These photographs should show the presence of subterranean water resources because vegetation growing above such water will remain green throughout the year and contrast sharply with the surrounding areas. In several areas containing bitumen (see Chapter 5), plants are stimulated into gigantism and luxuriant growth with a strong tendency to a second blooming and second vegetation. Areas showing luxuriant vegetation in the fall are therefore worthy of a ground follow-up.

In the steppe areas of West Kazakhstan, leguminous plants appear in the vegetation cover wherever phosphate beds are located close to the surface and can be readily distinguished by aerial color photography. The same authors have also observed that extensive areas of lichens and mosses occur in sandy deserts such as the Kara-Kum whenever gypsum deposits lie on or close to the surface. Absence of vegetation in aerial photographs should also be given close attention since this may be a measure of the presence of ultramafic areas.

6-3   AEROVISUAL SURVEYS

Although aerial photographs are extremely useful in the interpretation of plant cover, they do require supplementation by visual surveys carried out from aircraft at different altitudes. In the Soviet Union there is a state agency known as the All-Union Aerogeologic Trust (VAGT) which is responsible for aerial geologic and geobotanical mapping. Part of the standard work of this agency comprises a visual evaluation of terrain, usually after aerial photography has been carried out. Techniques of aerovisual survey have been well documented by Viktorov et al. (1964), and the use of these methods in arid regions has been described by Kasynova (1961) and Vostokova and Khdanova (1961). The following short account of procedures normally employed for this type of work is taken from Kasynova's work.

Aerovisual reconnaissance is carried out at the beginning of field work in order to give the geobotanist some idea of the vegetation and topography of the area to be studied. Flying over unknown territory, the observer makes notes both on aerial photographs and in a diary as he studies the vegetation and suitable access routes to the area to be investigated. Where details of the plant cover cannot be evaluated from the air, the geobotanist visits the doubtful areas on foot and completes his record of the vegetation in this way.

The next stage of the work is aerovisual mapping. This is best carried out by two observers, one of whom continually traces the route on aerial photographs and marks with appropriate symbols (see Chapter 3), the plant associations seen. The other observer logs each entry so that when the ground speed and flight course are known, the location of each entry can be pinpointed exactly. During the mapping procedures, the observer pays particular attention to features of the vegetation such as the color, pattern, and shades of the surface of the plant cover and also the color and form of the crowns of the trees. Flight altitudes should not exceed 600 ft during the mapping operations.

The final stage of the aerovisual survey is the coordination of all the hitherto accumulated data by flying a route over the area which will include all the major plant associations and geomorphological features. From this flight, a geobotanical profile of the whole area is prepared. Such an operation is the aerial analogue of a belt transect carried out on the ground.

Aerial geobotanical mapping from aerovisual surveys requires a group of very highly trained personnel, probably of different disciplines, working together in close collaboration. The Soviet Union is well ahead in this field, probably because so much of the work in the Western world is fragmented and carried on by many different agencies involving such diverse fields as agriculture, forestry, and geology. Many of these agencies do extremely good work—examples of this are the Australian Land Surveys published by C.S.I.R.O. (e.g., Mabbutt et al., 1963)—but it is very clear that a single agency responsible for geobotanical mapping and established in North America, Australia, or other appropriate areas, would be able to achieve far more than is presently being accomplished.

6-4   AERIAL INFRARED PHOTOGRAPHY

Perhaps the most spectacular advances in aerial photography have been in techniques for recording radiations in the near infrared part of the spectrum (7000–9000 Å). In geobotanical work, use is made of the fact that chlorophyll has a great capacity to reflect infrared radiation in the above spectral range and that this reflectivity may be recorded on film which is sensitive to this particular wavelength. Infrared photography may be carried out from the air either with black and white or with color film.

Figure 6-3 shows spectral reflectance curves for various types of vegetation. It will be noted that different types of foliage are not greatly different from each other in the visible part of the spectrum. The slight maximum at 550 nm (5500 Å) is all that is needed to give the characteristic green appearance of the vegetation. The infrared reflectances are, however, very different; and it is clear that a photographic emulsion sensitive to the infrared end of the spectrum should give much greater contrast than would a normal color film.

Some of the earliest infrared photography was that of Ives (1939) who showed from work carried out in Colorado that various types of vegetation could be easily

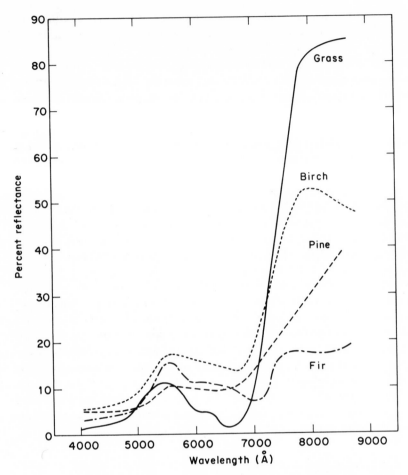

FIGURE 6-3 Spectral reflectance curves of various foliage types. From Fritz (1967). By courtesy of the American Society of Photogrammetry.

distinguished from the air with black and white film. Not only was it possible to distinguish more readily between different species of plants, but healthy and unhealthy specimens of the same species could also be differentiated, since the infrared reflectivity of chlorophyll diminishes very rapidly with ill health of the plant.

A typical infrared black and white film will have an emulsion that is sensitive to the whole of the visible part of the spectrum and to the near infrared part up to 9000 Å. Maximum sensitivity will be in the range 7700–8400 Å. It is customary to use a red filter with this type of film in order to absorb the blue end of the spectrum. The author's experience with this material in New Zealand is that better results can be obtained with infrared color film, and this will now be considered in detail.

Infrared color film (CIR) comprises three image layers which are sensitive to green, red, and infrared instead of blue, green, and red. A yellow filter (e.g.,

Wratten No. 12) is used to withhold blue light toward which the layers are also sensitive. Infrared film of this type has several disadvantages over normal color film. The first of these is that the exposure latitude is very limited ($\pm\frac{1}{2}$ stop) and unless exposures are exactly correct, disappointing results will ensue. Processing of the film will not normally be undertaken by most commercial interests and the services of a specialist have to be sought. Other disadvantages are that the infrared-sensitive image layer is somewhat unstable, has a short life, and must be kept cool at all times. Focusing is also a problem for close work since the positions for maximum focus are different for the infrared image and for the other two layers. In practice a compromise between the two has to be found. The author uses a camera with an attachment for infrared focusing and finds that overall this gives better results than focusing as for normal color film.

In spite of the above difficulties, good results can be obtained by relatively un-skilled personnel using conventional cameras operated from a light plane or heli-copter. It is advisable to set the camera at a suitable speed, for example $\frac{1}{500}$ second and for initial trials give exposures at values of $f16$, $f11$, $f8$, $f5.6$, $f4$, and all intermediate half stops. This is very wasteful of film, but since materials are so much cheaper than flying time, the waste is justified. As the operator gains experience, the number of exposures may be decreased to three or four in number, but to take fewer than this is risky.

The amount of infrared reflectance of vegetation decreases with altitude due to absorption of the radiation by water vapor in the atmosphere and also due to in-creased Rayleigh scattering which is especially pronounced at the blue end of the spectrum. Although the Wratten 12 filter normally used with CIR film cuts out wavelengths below 5200 Å where scattering is particularly serious, there is still sufficient interference remaining from this source to render infrared photography a problem at higher altitudes. For example, at 40,000 feet, as much as 90% of the infrared signal is lost.

Recently Pease and Bowden (1969) have discussed this attenuation problem and have shown that auxiliary filters with additional minus-blue filtration can greatly enhance the infrared signal from vegetation. For moderate enhancement, Kodak filters 82B or CC30B are recommended; for drastic enhancement, filters 80B, 80C, or CC50B may be used. The use of these auxiliary filters for high-altitude work does, however, present additional problems. For aerial work a shutter speed of at least $\frac{1}{500}$ sec is recommended. The filter factor for the CC50B filter is $1\frac{1}{2}$ stops and this involves working near the maximum of the aperture range of most cameras. In practice, it is sometimes necessary to compromise by reducing the shutter speed to $\frac{1}{250}$ sec and risk some loss of definition. Although the infrared signal is enhanced when auxiliary filters are used, there is less differentiation of vegetation in the image compared with photographs taken at a lower altitude with-out the use of these special filters.

The most obvious potential of remote sensing in the infrared part of the spectrum is in the preparation of geologic maps from consideration of the distribution of those plant species which are strongly affected by the geologic nature of the substrate.

Where mineralization toxic to plants is present in the soil, this should affect the vigor of the vegetation and may be evident in the photographs due to a difference in the infrared reflectivity of the chlorophyll of ailing vegetation. This procedure has already been used extensively in forestry and agriculture for detecting diseases in trees and crops (Norman and Fritz, 1965).

One condition for the successful use of the technique for detecting mineralization is that the vegetation cover should be so homogeneous that minor changes in individual specimens will be evident. In the author's own experience, this condition is seldom satisfied in New Zealand, where the very best that can be expected from the method is that it may be used to delineate boundaries between different geologic formations and not to detect mineralization within them.

Much of the pioneering work in infrared photography has been carried out in the United States (Colwell, 1964, 1967, 1968) mainly in connection with inventories of woodland resources and detection of disease in forest species. Colwell (1968) has described the *multiband camera* which has nine lenses and can photograph objects simultaneously with nine different types of film sensitive overall to the spectral range 3800–9000 Å. Multiband cameras can be carried in aircraft or satellites. Study of the nine photographs can enable the operator to obtain a *tone signature* for the object investigated and thus obtain more information than could possibly be gained from a single exposure.

6-5   AERIAL THERMOGRAPHY

The previous section has discussed photography in the range 7000–9000Å. It is not possible to devise photographic methods of photographing in the *thermal* part of the infrared spectrum since films for this purpose would have to be kept at very low temperatures before and during use and during processing. In practice it is possible to obtain indirect photographs of thermal emission by means of a so-called *optical–mechanical scanner* (Colwell, 1968). This device collects thermal emission from the ground by reflecting it with a system of mirrors onto a photoelectric detector which produces an electrical signal which is used to generate visible light. This light is recorded in turn on conventional black and white film and gives a permanent record of thermal emission from the object. Thermography would seem to have few applications in geobotany; its main uses are in military and fire prevention fields. It has nevertheless been included for the sake of completeness.

6-6   AERIAL RADAR IMAGERY

Remote sensing with radar has interesting applications in thickly forested terrain since it is capable of appearing to remove the vegetation cover completely and of exposing the underlying geologic and geomorphological features. Radar waves of long wavelength will readily penetrate vegetation and several feet of soil. Shorter wavelengths can be used to differentiate different species of vegetation.

In radar aerial imagery, a high-frequency signal is generated from the airplane and strikes the ground beneath. The reflected signal is analyzed by a special type of receiver, which, as with thermography, converts the energy into a visual form from which a conventional photograph is obtained.

Airborne facilities for radar imagery are now freely available. Wavelengths in the range of 1–10 cm should be useful for analyzing vegetation. Changes in frequency and phase of polarization of the reflected signal can be detected readily and used in identification of the terrain and its vegetation cover.

Much of the pioneering work on radar imagery applied to vegetation has been carried out by Morain and Simonett (1966). The technique may still be said to be in the development stage.

6-7   THE FUTURE OF AERIAL GEOBOTANICAL SURVEYING

The use of satellites for remote sensing has received a great deal of attention during the past decade. It is unfortunate that much of this attention stems from military considerations, but there is an increasing awareness among the "space nations" of the satellites' great potential for assessing natural resources. A satellite traveling at a height of 150 mi can survey an area of several thousand square miles at any one moment.

The advantages of remote sensing with a multiband camera have already been enumerated. There can be little doubt that if this is carried out from a satellite and combined with smaller-scale aerial surveys of the more likely areas, a very real contribution could be made to a geobotanical inventory of the whole of the world's resources.

SELECTED REFERENCES
See Table 6-1.

CHAPTER SEVEN
AN ASSESSMENT
OF
GEOBOTANICAL
PROSPECTING
METHODS

How effective are geobotanical methods of prospecting? This is a difficult question to answer since so much depends on the skill of the operator, the suitability of the area for this type of prospecting, and even on the nature of the criteria adopted to define success. Nevertheless, if case histories of successful use of the method are investigated, and if some consideration is given to probable trends in the future, it will be possible to form some sort of assessment of the potentialities of the technique and to place it in proper perspective with other methods of exploration.

A significant factor hindering a proper evaluation of geobotanical methods of prospecting is that much of this work has not been carried out by exploration companies but instead has been undertaken by organizations such as geological surveys or university groups often supported by these companies. A common pattern in this type of work is that the research organization develops the procedures by working in a region of known mineralization and then provides the data to an exploration company, which then presumably should apply the method in unknown areas where minerals are to be sought. In fact, this vital second step is seldom carried out (although there have been important exceptions to this rule). Although scientific knowledge has been enriched by the orientation survey, it becomes a piece of *pure* rather than *applied* research unless the mining company is itself willing to apply the technique or, at least, persuades the original workers to continue until the method has been applied in a number of unknown areas. Does the reluctance of exploration companies to apply what they have paid for stem from an innate conservatism, or does it arise from the fact that some geologists, secure and confident in their own specialty, become faltering and hesitant once the boundary into another scientific discipline has been crossed?

Almost without exception, successful use of the geobotanical method has resulted from thorough, *integrated* studies combining the initial survey with a subsequent exploration program and including other procedures such as biogeochemical, geochemical, and geophysical methods of exploration.

Another factor which can affect the success of an operation involves the selection of a suitable area for carrying it out. Clearly, a region completely devoid of vegetation would never be considered for geobotanical prospecting. On the other hand, there could well be areas that, although carrying extensive vegetation, would be no more promising, although this might not be obvious initially. Examples of this would be the presence of artifically introduced species growing over the area of the orientation survey but not elsewhere. Other unpromising regions would be where the vegetation cover was sparse and where the anomalies were small. In such cases, the minimum size of quadrat needed to define the vegetation assemblage might be appreciably larger than the anomaly, and geobotanical work would therefore be difficult and probably meaningless.

Subsequent sections of this chapter will be concerned with an assessment of two well-known case histories of successful use of the method; economics of the technique; its advantages and disadvantages; and finally, probable trends in the future.

7-1   SOME SUCCESSFUL USES OF GEOBOTANY IN MINERAL EXPLORATION

The well-known "copper flower" of Central Africa (*Ocimum homblei*— now *Becium homblei*), already mentioned in Chapter 4, and is probably one of the most reliable indicator plants ever recorded. The following extract from the original article describing its discovery (Horizon, 1959) will serve to show how successful its use has ultimately proved to be. The original work was undertaken by G. Woodward who was at the time (1949) employed by the Roan Selection Trust. Initially, the project nearly failed because *Ocimum homblei* was confused with the almost identical *Becium obovatum*, which is not a copper indicator and flowers at the same time. With practice, Woodward was ultimately able to differentiate between the two species and found that:

> The more information was collected, the more convinced some of the investigators became that the "copper flower" theory was fact. So faithfully did the flower follow the line of the deposits, the flower charts were almost identical in outline to the underlying copper deposits. Where the ore bodies narrowed, so did the distribution of the flowers and where the ore widened out in folds, as at Roan Extension, Muliashi, the floral spread not only followed suit but appeared in greater density. Over known deposits, the abruptness with which it ceased growing at almost the precise edge of the mineralization, showed that here indeed was a sensitive copper indicator of remarkable selectivity.... By 1954, observations had been made at 30 widespread copper occurrences from near Lusaka to the Congo border. In all but two cases, the faithful floral signposts were there. In both exceptions the soils over the mineralized formations were deeply leached and had a low copper content.... It is now nine years since Copperbelt Geologists' researches began and in that time they have more than once had the satisfaction of seeing the occurrence of *Ocimum homblei* in places where copper was not previously known, lead to soil sampling, pitting and finally drilling with successful results, ... pits were being sunk on a long-known ore body.

> About three-quarters of a mile from its then known limits, an extensive growth of the copper flower had been previously recorded on a floral map. ... Drill rigs came in and among the sage clumps that had brought them there, the diamond bits went down to intersect large new reserves.

In follow-up work on *Becium homblei*, Howard-Williams (1970) considered that although this plant is apparently a copper indicator, this is indirectly so and is due to competition from the related *B. obovatum* which is intolerant of copper-rich soils. *Becium homblei*, being copper tolerant, can only survive where its competitor cannot grow.

The work of H. L. Cannon and her colleagues on the Colorado Plateau has already been mentioned (see Chapters 3 and 4), but it would be appropriate at this stage to summarize the results of their work when applied to unknown areas. In operations carried out in 6 sq mi of the Yellow Cat Area, Grand County, Utah (Cannon, 1964), 5 uranium ore bodies were found by the use of plants alone. Another striking feature of this work was that some ore bodies were found in ground which would not normally have been assessed as being geologically favorable if indicator plants had not been present. Two ore bodies were also found in the Poison Canyon area of the Grants District, New Mexico (Cannon, 1960a), again by the use of plants only.

7-2   THE ECONOMICS OF GEOBOTANICAL METHODS OF PROSPECTING

It is not easy to give a general evaluation of the economics of geobotanical methods of prospecting since these techniques are themselves so diverse. In the case of those studies involving work on the ground, the execution is simple and the economics of the application of the method involve only the cost of mapping plant distributions. For this part of the work, the costs must surely be far less than for any other exploration method not involving an aerial survey. Not only does the material not have to be sampled, but neither does it have to be analyzed, and the results are available immediately. Unfortunately, however, the cost of the preliminary orientation survey has to be added to the overall cost of the whole project and this is a sum which cannot be assessed except in individual cases. In general, the shorter the orientation survey and the greater the amount of accessible virgin territory which has to be investigated, the more economic geobotanical methods become. Geochemical methods such as sampling of soils and stream sediments do not require such an extensive orientation survey and are probably more economic when small areas are to be covered.

Too much emphasis should not, however, be placed on the cost of the orientation survey in geobotanical projects since once a *universal indicator* has been detected, it may be used presumably in any area without the need of further preliminary work.

The economics of geobotanical work carried out from the air are very different from those of ground operations. From the author's own experiences in New Zealand, aerial photography, particularly in the infrared region of the spectrum, can produce far more information than ground surveys and at a cost that is not prohibitive. In such cases, the orientation survey and applications of the method to virgin territory can be carried out simultaneously by using infrared photography to study vegetation over known mineralization and over the unknown areas. Aerial infrared photography (see Chapter 6) alone is usually sufficient to carry out belt transects of the tree layer (though not of the shrub layer) of the vegetation cover with very little supporting ground work. The data obtained by such methods are, however, usually less reliable than those obtained by ground operations. The procedure is most useful when comparatively large areas are involved. For further discussion of costs in geobotanical prospecting, see Hawkes and Webb (1962, p. 321).

7-3   ADVANTAGES AND DISADVANTAGES OF GEOBOTANICAL PROSPECTING METHODS

A number of advantages and disadvantages of geobotanical methods of prospecting are listed below.

ADVANTAGES
1. Once an orientation survey has been carried out, costs are extremely low.
2. Different geologic formations as well as mineralization within them can be detected by geobotanical observations.
3. Aerial methods can be applied to the procedure with consequent saving in time and effort.
4. Indicator plants can sometimes show the presence of mineralization at depth under conditions where other methods would give a negative response.

DISADVANTAGES
1. A high degree of individual skill is needed from workers in this field.
2. Data obtained from orientation surveys are not necessarily of universal application and may have only local significance.
3. In some cases the method can only be applied seasonally, as for example, when plants are in flower.
4. The method may only be applied where vegetation conditions are favorable.
5. In the West there is little coordinated research in this field being undertaken at present.

The above listing of advantages and disadvantages is not of course complete but the main points have been covered. A consideration of paramount importance is the fact that geobotanical prospecting is just one of many techniques and is meant to supplement rather than replace other methods. Until a sufficiently large pool of skilled workers in this field can be trained, it will not be possible to test this procedure adequately or to compare it objectively with other methods of exploration.

7-4   FUTURE TRENDS IN GEOBOTANICAL PROSPECTING

There can be little doubt that major advances in geobotanical prospecting in the future will be centered around the technique of remote sensing either from aircraft or from satellites. The great advantage of aerial observation lies in the vast amount of information which can be collected and recorded in a short space of time. Although the technology of *interpretation* is lagging some way behind, there is a real possibility that this bottleneck will be overcome in the near future as computerized techniques become available.

Vinogradov (1963) has described a microphotometric method for scanning aerial photographs. The different density readings corresponding to different types of vegetation are plotted as a function of the spacing along the scanning direction. By use of various parameters, the nature and abundance of the vegetation is calculated.

Rommel et al. (1968) have suggested the possibility of using an airborne multichannel optical-mechanical scanner (designed by Tandgreke and Phillips) to obtain the *signature* of each species of vegetation in the survey area. The instrument works on the principle of a slit image that is scanned by a rotating mirror to bring the whole field of view into use. The beam is split into its component wavelength bands by means of a prism or grating and the intensity of each component is measured by a detector. The outputs of the detector are recorded on a multichannel tape recorder or can be fed into a computer. The final recording is in fact the characteristic spectrum of each type of vegetation being a function of its reflectivity at various wavelengths. Applications so far have been agricultural, but there is no reason why the procedure should not be used in the search for minerals by geobotanical means.

Radar imagery also seems to have some potential for the future. Further discussion of this subject is rather outside the scope of the present work and the reader is referred to Rouse et al. (1966) for a fuller treatment.

Increased use of computers will also be made in the future for the processing of data. Even today, relatively simple statistical calculations involving the relationship between plant distributions and mineralization in the substrata are impracticable for large numbers of samples without the use of the computer.

It might at first seem to be difficult to predict what future improvements might be made in field studies of indicator plants and indicator-plant communities since such operations tend to be visual and do not lend themselves to automation. However, if efficient instruments could be designed for in situ measurements of elemental concentrations in soils and rocks, this will be immensely useful in orientation surveys since at present the geobotanist has to rely on chemical analysis of the substrate for a thorough study of indicator plants. The development of gamma-source X-ray fluorescence units which are both portable and inexpensive is beginning to make such in situ observations a practical possibility. The instruments do not at present give results which are particularly reproducible and they are not very sensitive. Nevertheless in their present form they will be more than adequate as a rough monitoring device to enable the geobotanist to decide whether he is working on mineralized or nonmineralized ground. X-ray fluoresence units will not in themselves replace geobotanical methods since, even under the best conditions, they can never hope to compete in speed with operations that are purely visual in their ultimate applications.

Despite an appreciable interest in geobotanical methods in the West during the 1950s, interest has waned somewhat in recent years. This must surely be a transient phase since the potentialities of remote sensing alone should do much to increase interest in the years to come. There is, however, another factor which cannot be ignored. In time, and probably in the foreseeable future, a very large part of the world's surface will have been surveyed by various geochemical and geophysical prospecting techniques. Geologists will have left until the last what is to them, the least familiar of potential methods. When that time comes, geobotanical methods of prospecting will be the last of potential techniques to be exploited to the full in the search for minerals which will become progressively more elusive with each passing year.

PART TWO
BIOGEOCHEMISTRY
IN
MINERAL
EXPLORATION

CHAPTER EIGHT
INTRODUCTION
TO
BIOGEOCHEMISTRY
IN MINERAL
EXPLORATION

Biogeochemical methods of prospecting depend on the chemical analysis of elements in vegetation. The assumption is made that an element in the soil or bedrock will be accumulated by the plant in a reproducible manner and that consequently, anomalous amounts in the vegetation will indicate anomalies in the substrate.

The accumulation of certain elements by humus and vegetation was reported by Goldschmidt and Vernadsky over 40 years ago but it was not until about 10 years later that the biogeochemical method began to evolve slowly from these basic principles. The technique is very much younger than geobotany, since it depends on the existence of simple, rapid, sensitive, and inexpensive methods of analysis which did not become available until the work of Goldschmidt and his co-workers in the 1930s on emission spectrography.

8-1 HISTORY OF THE BIOGEOCHEMICAL METHOD

Pioneering work on biogeochemical prospecting began independently in Scandinavia, England, and the Soviet Union in the years 1938 and 1939. Credit for the development of the method is usually assigned to Brundin or Tkalich.

Brundin (1939) undertook biogeochemical investigations on vanadium in Sweden and tungsten in Cornwall and took out a patent for the method.

Tkalich (1938) simultaneously carried out biogeochemical investigations at the Unashinsky arsenopyrite deposits of Eastern Siberia and discovered that the ore body could be delineated by the iron content of the local vegetation.

During World War II and its immediate aftermath, Scandinavian workers were active in developing biogeochemical methods in Norway (Vogt, 1939, 1942a, b; Vogt and Bergh, 1947, 1948; Vogt and Braadlie, 1942; Vogt and Bugge, 1943), Finland (Rankama, 1940, 1941), and Greece (Hedström and Nordström, 1945).

The breakdown of scientific communication during the war years led to other workers in the field being unaware of the pioneering work of Brundin, Tkalich, and Vogt. The most successful and extensive of these early developments was in British Columbia where Warren and his co-workers began their investigations as early as 1944. The history of this work has been recorded by Warren and Delavault (1950b) who tell of how, in 1944 when the senior author was engaged in the removal of 50,000 tons of overburden from an area of possible mineralization, "it was impossible to overlook the way in which the roots of trees and various lesser plants penetrated into the very places which we were reaching with so much effort." In 1945 Warren and his co-workers began field operations which showed that the copper and zinc content of some trees and lesser plants could reflect to some extent the presence of these elements in the underlying soils or rock formations.

From 1948 onward, the pace of development of the method in British Columbia accelerated considerably. Thus, by 1949, Warren was able to report that, "the reaction to biogeochemistry in British Columbia, had changed from one of overall disbelief to benevolent skepticism." In the period 1947 to 1966, Warren and his co-workers published some 27 papers concerning the development of the biogeochemical method (Warren, 1962, 1966; Warren and Delavault, 1948, 1949, 1950a, b, 1951, 1952, 1955a, b, c, 1957, 1959, 1960, 1965; Warren, Delavault, and Barakso, 1964, 1966; Warren, Delavault, and Cross, 1959, 1966a, b; Warren, Delavault, and Fortescue, 1955; Warren, Delavault, and Irish, 1949, 1951, 1952a, b; Warren, Delavault, and Routley, 1953; Warren and Howatson, 1947).

The classical work of Warren and his co-workers has done much to lay the foundations for biogeochemical prospecting as a respectable exploration tool. In light of the fact that this work was undertaken before the advent of atomic-absorption spectrophotometry (see Chapter 15) and the computer (see Chapter 13), the progress made is all the more remarkable.

The biogeochemical method was refined and widely used in unknown areas by the U.S. Geological Survey in the 1950s and early 1960s. One of the first to apply the method in the United States was Harbaugh (1950), who investigated base metals in soils and vegetation of the Tri-State region.

The work undertaken by the U.S. Geological Survey was carried out on the Colorado Plateau as part of a program of exploration for uranium. This work was undertaken mainly by Cannon and her co-workers (Cannon, 1957, 1960a, 1964) and represents a most thorough and successful application of the method—as a result of which, several uranium ore bodies were discovered.

Following the pioneering work of Tkalich (1938, 1952), Russia has remained the main center of biogeochemical prospecting work in the world. After the interruption caused by World War II, research into biogeochemical methods was resumed in 1948 on a large scale, although a geobotanist was included in all major Soviet geological expeditions after 1945.

Much of the later Russian work was carried on by Malyuga and his co-workers at the Vernadsky Institute, Moscow (Malyuga, 1947, 1950, 1951, 1954, 1958a, b, 1959, 1960; Malyuga et al., 1959, 1960; Malyuga and Makarova (1955); Malyuga and Petrunina, 1961).

8-2   BIOGEOCHEMICAL TOPICS COVERED IN THE PRESENT WORK

The successful utilization of biogeochemical methods of prospecting involves a knowledge of many disciplines, including botany, chemistry, ecology, geology, plant chemistry, plant physiology, soil science, and statistics. It is for this reason, more than any other, that the development of the method has been slower than that of other methods such as geophysics. In Part Two an attempt has been made to bring together the basic elements of most of the above disciplines insofar as they apply to biogeochemical prospecting. It is hoped that this work will be instrumental to some small degree in helping to bridge the gap between the field worker (who should have some background knowledge of these subjects) and the various specialists. Lack of communication between the two is a handicap to further development of the technique, and until this is overcome, the full potential of biogeochemical prospecting will not be realized.

Topics discussed in Part Two will therefore include the formation of soils and the distribution of the elements in them, elemental accumulation by plants, and a practical guide to field work for orientation and exploration surveys. There will also be discussion of a statistical approach to biogeochemical prospecting, as well as a statistically orientated treatment of the problem of essential and nonessential trace elements in mineral exploration. A chapter on the chemical analysis of plants and soils will attempt to bridge the gap between the field worker and analyst and is followed by case histories involving biogeochemical prospecting in rivers, peatlands, and rain forests.

It is not claimed that biogeochemical prospecting will be a universal panacea for the prospector in his search for minerals; under some conditions it is unwise to expect too much from the method. If used properly, however, by persons versed in its potential, the method will surely find an increasing area of application in the search for minerals in all parts of the world.

SELECTED REFERENCES
History of biogeochemical prospecting: Warren and Delavault (1950b); Malyuga (1964)

CHAPTER NINE
SOILS
AND SOIL
FORMATION

Volcanic action and the disintegration and decomposition of large masses of solid rock have left a blanket of unconsolidated material on the surface of the earth. This layer of mineral material is known as the *regolith* and contains varying amounts of organic material from the decomposition of living organisms growing upon it.

Soil forms the upper layered part of the regolith and is differentiated into *horizons* of varying depths differing morphologically and chemically from the parent material.

A basic knowledge of soils and soil formation processes is an essential prerequisite for the proper understanding of biogeochemical prospecting techniques. This knowledge is particularly necessary since vegetation and the soil are so intimately connected and interdependent.

The science of *pedology* has developed only comparatively recently and is still relatively young compared with other natural sciences. The neglect of pedology and soil science in general has historical reasons. Before the nineteenth century, the soil was regarded as being as lowly as the humble peasant or serf who tilled it and was considered by most scientists as being less worthy of study than "more interesting" subjects such as physics, chemistry, and mathematics. This neglect was encouraged by a second factor and this was the fact that in Western Europe, at least, so much of the readily accessible land was already under the plow that little undisturbed soil remained. It is not surprising therefore that the development of soil science began in Russia, where there was still virgin soil.

The great Russian pedologist Dokuchaev (1846–1903) is usually credited as being the founder of soil science. In his famous book *The Russian Chernozem*, Dokuchaev (1883) noted that *chernozems* (heavy black soils) were always associated with grassland vegetation and he considered them to be intimately related. This theory was in marked contrast to those of classical geologists who had always considered that chernozems belonged to a particular geological formation. Dokuchaev also studied the *podzols* of forested areas and noticed the various horizons found in these soils. He laid the foundations of the *zonal* theory of soils, which established that soil formation is strongly related to climate.

Another great Russian soil scientist, Glinka (1867–1927), a pupil of Dokuchaev, was responsible for the organization of the subject of soil science and offered the world's first course in soil science at the University of St. Petersburg in 1911.

Part of the relative popularity of biogeochemical methods of prospecting in the Soviet Union springs from the preeminence of this country in soil science and from the existence of soil maps of a large part of the country, prepared in many cases over forty years ago.

Soil science is now a well-established subject in most countries of the world and is receiving ever more attention as a useful discipline for increasing the efficiency of agriculture to feed the world's growing population. This recognition of the great value of soil science does not always apply to workers in other fields. Hawkes and Webb (1962) have commented that: "It is perhaps surprising that geologists in general have little understanding of the soils which mantle so much of the earth's surface since any information that can be gained from soils concerning the bedrock geology must be of value." With the increasing use of soil sampling in geochemical exploration, a proper understanding of the nature of soils and of the factors involved in their formation is becoming ever more important.

The purpose of this chapter is to give a short introduction to the subject of soils and soil formation with particular emphasis on the biologic principles involved in *profile* development. In the Soviet Union soil sampling, particularly of the humic layer, is considered to be a part of biogeochemical prospecting; and in view of the interdependence of soils and vegetation, this definition is probably justified. In this present work, however, a clear distinction between the two is made, particularly as soil sampling has been more than adequately covered by Hawkes and Webb (1962). Soil sampling will only be discussed insofar as it is an adjunct to prospecting methods involving plant analysis.

9-1   SOIL HORIZONS AND SOIL PROFILES

The sequential layers in properly formed soils are known as *horizons*; they form a progression from the upper humic layer down to the weathered parent material. The whole sequence is known as a soil *profile*.

If the surface covering of loose organic litter is ignored, the soil profile may be said to consist of three main horizons. The upper layer known as the *A horizon* is usually *eluvial* in nature where leaching has removed soluble salts and material fine enough to form a suspension. The eluviated material is deposited in the lower *B horizon* which is hence essentially *illuvial* in character. These two horizons together form the *solum* and are differentiated from the lower *C horizon*, which is essentially weathered parent material and is uniform throughout. The C horizon can also comprise old soils from which new soils are formed by a process known as *polygenesis*. The eluviated constituents are usually clay minerals, iron, humus, and lime. Figure 9-1 shows a typical series of profiles for various soil types.

The broad designation of three main horizons is usually insufficient to define

a.

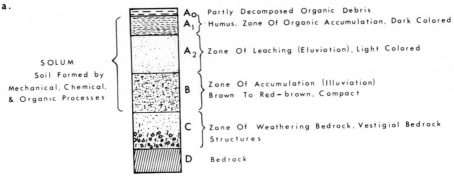

SOLUM
Soil Formed by
Mechanical, Chemical,
& Organic Processes

A$_0$ — Partly Decomposed Organic Debris
A$_1$ — Humus, Zone Of Organic Accumulation, Dark Colored
A$_2$ — Zone Of Leaching (Eluviation), Light Colored
B — Zone Of Accumulation (Illuviation) Brown To Red-brown, Compact
C — Zone Of Weathering Bedrock, Vestigial Bedrock Structures
D — Bedrock

b.

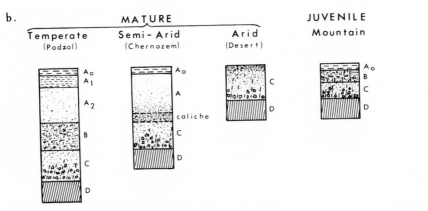

MATURE   JUVENILE
Temperate   Semi-Arid   Arid   Mountain
(Podzol)   (Chernozem)   (Desert)

FIGURE 9-1 Diagrammatic representation of soil profiles. (a) Typical soil profile. (b) Climatic variations. From Andrews-Jones (1968). By courtesy of the Colorado School of Mines.

completely all the various layers that may exist in the soil. A number of different conventions have been used to describe these subhorizons, for example, designations such as A$_0$, A$_1$, A$_2$, ..., B$_1$, B$_2$, ..., etc. Until recently there were no international standards on nomenclature and a great deal of confusion has arisen. In 1968, at the 9th International Conference on Soil Science held at Adelaide, draft proposals for an international system of soil nomenclature were discussed but have yet to be finalized. The main provisions of the new system have been described by Corbett (1969). The use of A, B, and C for the eluvial, illuvial, and parent horizons has been retained; but a number of other letters have been incorporated. These are abbreviations of well-known words and are easy to memorize. Table 9-1 shows in a simple form the proposed international system. It will be noted that the classification includes the leaf litter at the top and the bedrock beneath. It is seldom necessary in geochemical or biogeochemical prospecting to differentiate between horizons to the extent shown in this table, but a knowledge of the international

TABLE 9-1   Proposed International System for Description of Soil Horizons

Organic mat	Ol	Fresh leaf litter
	Of	Partly decomposed leaf litter
	Oh	Well-decomposed leaf litter
	OA	Transition organic mat and first mineral horizon
Eluvial horizons	Ah	Weakly leached horizon with humus accumulation
	E	Strongly leached horizon light in color
Illuvial horizons	Bh	Strong humus enrichment
	Bfe	Strong iron enrichment
	Bfe,g	Strong iron enrichment with mottling due to water
	CB	Transition between illuvial and parent horizons
Weathered parent material	C	Altered rock with structure preserved
Bedrock	R	Fresh rock

SOURCE: Corbett (1969).

system will be useful in interpretation of the relevant literature. Note that all horizons comprise the regolith, whereas the solum does not include bedrock and the weathered parent material.

9-2   FACTORS IN SOIL FORMATION

The classic work of Dokuchaev revealed that five factors are responsible for the formation of soils: the nature of the parent material; climate; topography; the age of the developing profile; and biologic activity. Many of these factors are interdependent and in the following discussion it is assumed that when any one of these factors is considered, the other four are controlled. Factors of soil formation have been discussed by Jenny (1941) and by Corbett (1969). Much of the material in this chapter is derived from these two sources.

THE EFFECT OF CLIMATE   Climate has a dominant influence on the development of most soils. Such soils are said to be *zonal*, whereas *intrazonal* soils are those dominantly affected by factors other than climate. A third class includes soils too young to have achieved much differentiation of the parent material. These are said to be *azonal* and are strongly influenced by the time factor.

Figure 9-2a shows in simplified form the main effects of climate on soil development. (Figure 9-2b will be discussed later.) Under hot, wet conditions, humic material tends to be oxidized as fast as it is formed and therefore *laterites* and *krasnozems* lack the black and brown pigmentation found in soils such as podzols which are formed under cooler conditions.

Under cold conditions, organic matter accumulates much faster than it is oxidized and peat and peat bogs tend to form.

Under desert conditions, there is not sufficient organic material to darken the soils and they assume the color (red) of one of their major constituents, ferric oxide.

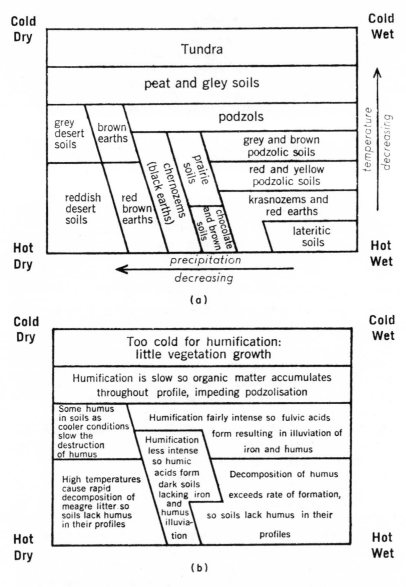

FIGURE 9-2 (a) Soil zonation in response to climate. (b) The role of organic matter in soil formation in response to climate. Boundaries in (b) correspond to those in (a). From Corbett (1969). By courtesy of Martindale Press Pty. Ltd.

In tropical soils, there is an increased decomposition of primary minerals into secondary minerals such as clays, due to the effects of heat and moisture. The high content of colloidal clay in these soils results in little illuviation so texture differentiation is slight. In desert soils, illuviation is again slight, due to the lack of rainfall.

In podzols the colloidal content is less than in tropical soils so that, despite appreciable rainfall, illuviation can readily occur and there is good differentiation within the soil profile.

The pH and humus content of soils are both a function of rainfall. Since humus gives an acidic reaction, peats and podzols usually have a low pH. Laterites and desert soils with their low humus content have relatively high pH values. The high pH of desert soils is further encouraged by their frequent content of basic sodium and magnesium salts which are not leached due to lack of rainfall. The relationship between rainfall and pH is illustrated by Figure 9-3 which shows the close concordance of rainfall and soil pH maps of the state of California.

THE EFFECT OF THE PARENT MATERIAL   Although it might be expected that the nature of the parent material would be the dominant factor in soil formation, in fact this is often subordinate to climate unless the soils are *juvenile*. Soil structure is to some extent affected by the nature of the bedrock, and the rate of soil formation increases with rock types which are readily weathered. Where soils are formed from a rock parent material rather than from preweathered material, the soil tends to be shallow.

Soil color is seldom a function of the parent material and is essentially controlled by climatic factors. One soil factor that is fairly closely controlled by the nature of the parent material is the abundance of the major and minor elemental constituents.

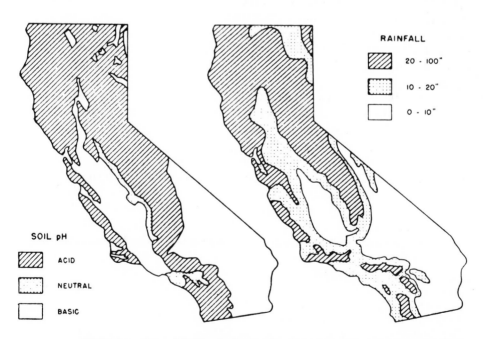

FIGURE 9-3 Maps of California showing the close relationship between soil pH and the amount of rainfall. Modified by Carlisle and Cleveland (1958) from Jenny (1951). By courtesy of the California Division of Mines and Geology.

Indeed, this is the very factor which facilitates geochemical prospecting by soil sampling.

THE EFFECT OF RELIEF   The effect of slope is to produce a localized microclimate and to cause significant differences in profile development. In general, the steeper the slope, the shallower the soil and the less differentiation in the soil profile. Water moving down the slope leaches out easily mobilized constituents and precipitates them downhill in areas of gentle relief. For this reason the lower slopes tend to have a higher pH, as it is here that lime is deposited.

In low-lying areas, there is less free drainage and the water table may lie almost on the surface. These soils tend to have a higher organic content than those that are well drained. Peat is the ultimate development of this trend.

THE EFFECT OF TIME   The time taken for a well developed soil profile to appear is largely a function of climate. Under hot, wet conditions, a soil profile will develop much faster than in a cold, dry climate. Quantitative measurements of rates of soil formation have been rendered possible by studying new geological formations formed after natural disasters or by observing soil development on moraines of retreating glaciers.

The volcanic explosion of the island of Krakatoa in 1883 afforded an excellent opportunity to study the rate of soil formation on freshly weathered material. After 45 years, a layer 14 in. deep was observed to be fairly well differentiated from the parent material in conditions which were admittedly favorable because of the tropical climate. By contrast, Chandler (1942) studied the rate of podzolisation of a moraine from the retreating Mendenhall Glacier near Juneau in Alaska and found that 500 to 1000 years were needed for a well-developed soil profile to appear.

Soils are sometimes classified as *juvenile* or *mature* depending on their state of development. In steep areas with high rates of erosion, soils can remain juvenile permanently if the rate of erosion is faster than the rate of formation.

THE EFFECT OF BIOLOGIC ACTIVITY   The formation of a soil profile and the development of a concomitant vegetation are so closely linked that it is often impossible to say which came first. When fresh rock is exposed to weathering, the first plants (usually lichens and mosses) colonize the surface and both the vegetation and soil increase in complexity with the passing years. As the humus content of the soil increases, so does its water-holding capacity with the result that higher plants can then survive in the new environment. The concomitant development of vegetation and soil is known as *seral progression* and terminates with the so-called *climax* vegetation and mature soil. Seral progression takes place much more quickly in tropical regions than in cold, dry areas.

The Krakatoa explosion also provided an excellent opportunity for ecologists and soil scientists to study seral progression. A few years after the explosion, grasses had become reestablished on the ash. By 1900, a savannah-type drought-resistant shrub layer had appeared. Later, forest trees became established on the soil profile; today the vegetation is similar to that existing before 1883, although somewhat poorer in species.

The accumulation of humus in soils is largely a function of climate as mentioned above. The distribution of organic matter in the soil is controlled by the rainfall and permeability of the soil. Thick chernozems, found under semiarid grasslands, have a uniform distribution of humus through the A and B horizons, whereas heavily leached podzols have discrete zones of organic material. Figure 9-4 shows the distribution of humus in zonal soils of the Soviet Union. The contrast between thick chernozems and podzols is particularly striking.

Humus contains a large number of colloidal organic compounds which are grouped under the headings of *humic* and *fulvic* acids. These two components differ greatly in their mobility and effect on the soil. The fulvic acids assist in the mobilization of various elements, particularly iron and aluminum. Marked illuviation will occur when these acids are present. Humic acids by contrast are far more stable and tend to impede illuviation. Table 9-2 gives the composition of humus in some principal soil groups of the Soviet Union. The ratio of humic to fulvic acids varies appreciably among different soil types. In chernozems, the increased amount of humic acids results in less leaching of metals in the soil profile and therefore

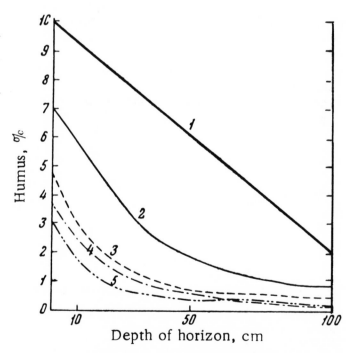

FIGURE 9-4 Distribution of humus in zonal soils to depths of 100 cm. 1, thick chernozems; 2, ordinary chernozems; 3, grey forest soils; 4, chestnut soils; 5, podzols. From Malyuga (1964). By courtesy of the Plenum Publishing Corp.

TABLE 9-2   Composition of Humus in Zonal Soils of the Soviet Union

Soil	Mean humus content (%)	Humic acids % of humus	Fulvic acids % of humus	Insoluble % of humus	Ratio humic to fulvic acids
Podzol	3.5	15–25	47	28	0.4
Thick chernozem	10.0	40	39	19	1.0
Ordinary chernozem	7.5	35	37	25	1.0
Dark chestnut	3.5	34	35	26	1.0
Grey soils	1.5	21	41	32	0.5
Red soils	5.0	15	50	33	0.3

SOURCE: Malyuga (1964).

less differentiation. In podzols, by contrast, the increased amount of fulvic acids encourages illuviation.

The mass of organic material produced per annum is very much a function of the climate and type of vegetation. As might be expected, tundra and desert associations are the least productive; tropical rain forests produce the greatest amounts. Table 9-3 lists the production of organic matter for different types of vegetation in different climates.

The pH of humic material is to some extent a function of the type of vegetation that has produced it. Grasses tend to have a higher calcium content than woodland species with the result that calcium humates, which are basic in reaction, are produced to a greater extent.

Figure 9-2b shows the role of organic matter in soil formation in response to climate. The soil boundaries are the same as in Figure 9-2a.

Figure 9-5 shows the gradation of soil types under the influence of climate and relief and summarizes pictorially some of the above discussion on soil formation.

TABLE 9-3   Production of Organic Matter by Vegetation

Type of vegetation	Annual production (tons/acre)
Alpine meadows	0.22–0.40
Short-grass prairie	0.71
Tall-grass prairie	0.85–1.73
Mixed tall-grass prairie	2.22
Pine forest	1.42
Beech forest	1.46
White-pine needles	2.09
Beech-birch-aspen forest	2.85
Tropical primeval forest	11.10
Tropical savannahs	13.30
Monsoon forest	22.20
Tropical legumes	24.40
Tropical rain forest	45–90

SOURCE: Jenny (1941).

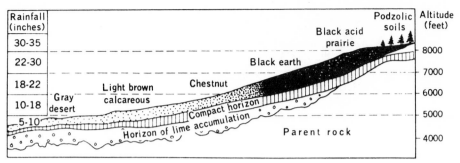

N.B. Scale of soil profiles is greatly exaggerated

FIGURE 9-5 Gradation of soil types from desert to humid mountain top, west slope of Big Horns, Wyoming. Modified by Hawkes and Webb (1962) from Thorp (1931).

9-3   CLASSIFICATION OF SOILS

Classification of soils is extremely complicated because so many different systems are in current use. Engineers and architects use the *taxonomic* system, which is based largely on the physical properties of soil, such as shear strength, cohesiveness, texture, etc. The *morphological* approach takes into account total profile features and reflects the soil-forming processes. The *genetic* system (developed by Dokuchaev and his co-workers) is related to the five factors of soil formation. Modern soil classification commonly employs elements of all three systems.

The primary classification of soils is the *soil order*, which usually relates to the five factors of soil formation: zonal (climate); intrazonal (relief, vegetation, and parent material); azonal (time). *Soil suborders* are usually associated with the main profile features such as domination by organic matter, marked illuviation, etc. The third classification is that of the *great soil groups* and is based on the total profile features. Typical nomenclatures of the great soil groups are: red podzols, chernozems, solonchaks, etc. This classification is analagous to the common names of rocks as used by geologists (e.g., granites, andesites, basalts, etc.).

There are at present at least five commonly used systems of soil classification (Smith, 1960; Stephens, 1962; Northcote, 1965; Stace and Hubble, 1968; FAO/UNESCO, 1968). The FAO/UNESCO system is an attempt at internationalization and is to be used for a 1:5,000,000 soil map of the world. The terminology used in this system is radically different from that used in other classifications and will not be used in this present work since most of the existing literature on soil science uses the other systems.

SELECTED REFERENCES
General: Jenny (1941); Corbett (1969)

CHAPTER TEN
THE DISTRIBUTION AND MOBILIZATION OF MINOR ELEMENTS IN SOILS

In Chapter 9 the nature and formation of soils were discussed without consideration of the minor (trace) constituents (concentrations less than 0.1% in the soil) of the material. The distribution and mobilization of these trace elements in soils will now be considered in order to lay a foundation for the discussion of plant nutrition and the mechanisms whereby elements may be accumulated by vegetation.

10-1   THE DISTRIBUTION OF THE MINOR ELEMENTS IN SOILS

Distribution of the minor elements is a function not only of soil type but also of particle size and depth within the profile. The nature of the parent material determines the abundances of the elements in the profile as a whole but the mobility of the constituents, together with other factors such as climate, determine their distribution within the soil profile. Table 10-1 lists the average abundances of trace elements in soils, the crust of the earth, sediments, and igneous rocks. The range of values normally encountered in soils is shown in Figure 10-1.

Table 10-1 and Figure 10-1 are useful indicators of the threshold values that might be expected in areas not containing mineralization. Some elements, such as zinc, have a relatively uniform concentration in most soils and rocks; other elements, such as chromium, cobalt, and nickel, are heavily dependent on the nature of the bedrock.

In most soils the concentrations of minor elements in various horizons of the profiles are far from constant. Table 10-2 shows the distribution of a number of elements within profiles of five common zonal soils collected during a nationwide soil survey of the Soviet Union. All profiles except for those of gray desert soils, show an accumulation of all five elements in the upper humic horizon. In all cases, with the same exception, there is relative depletion of the elements in the A_2 and B_1 horizons due to leaching to the lower layers. The desert soils show depletion of

TABLE 10-1   Average Abundances of Trace Elements in Soils, the Earth's Crust, Sediments, and Igneous Rocks

Element	Soil (ppm)	Crust (ppm)	Sediments (ppm)	Igneous rocks (ppm)
Ti	4600	5000	4500	4400
Mn	850	975	760	1000
Ba	500	450	690	640
Zr	300	190	200	170
Sr	300	385	450	350
Cr	200	150	130	117
V	100	145	130	90
Rb	80	165	270	280
Zn	50	125	80	80
Ce	50	46	50	40
Ni	40	95	95	100
Li	30	45	60	50
Ga	20	15	30	20
Cu	20	75	57	70
Nb	15	20	20	20
Pb	10	15	20	16
Sn	10	40	16	32
B	10	10	56	13
Co	10	35	22	18
Th	13	10	11	13
Be	6	4.5	5	4.2
Ge	5	4	4.5	2
As	5	3.4	6.6	2
Cs	5	1	10	10
Mo	2.5	2.3	2	1.7
U	1	2.4	3.2	2.6
Ag	1	0.06	0.5	0.2
Cd	0.5	0.2	0.5	0.13
Se	0.5	0.07	0.6	0.01
W	—	1.2	2	2
Sb	—	0.6	1.25	0.3
Hg	0.01	0.07	0.04	0.06
Bi	—	0.34	0.01	0.1
Au	—	0.001	—	0.001

SOURCE: Andrews-Jones (1968).

the A horizon probably due to the lack of humus and clay minerals in this part of the profile.

The uneven distribution of trace elements within a soil profile highlights the care which must be taken during geochemical prospecting to ensure that the correct horizon is sampled. This factor is further emphasized in Table 10-3, which shows the horizons in which maximum enrichment occurs for a number of elements in various soil types.

Trace elements are unevenly distributed within a single horizon if consideration is given to particle size. Figure 10-2 shows the distribution of nickel in soil fractions of different mesh size and demonstrates the great differences that can exist. Other elements, for example tin as cassiterite, show the same tendency. As an example, Williams (1934) has shown that 98% of the cassiterite in soils and sediments of Stewart Island, New Zealand is concentrated in the +20 fraction.

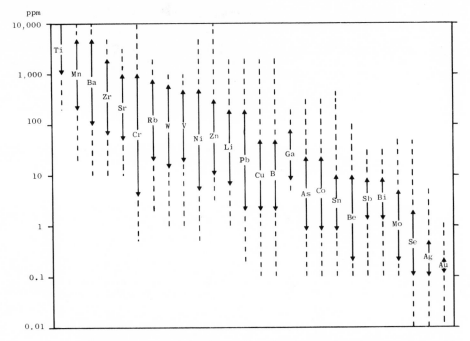

FIGURE 10-1 Range of abundance values of some trace elements commonly found in soils. Dashed lines indicate the more extreme values. From Andrews-Jones (1968). By courtesy of the Colorado School of Mines.

TABLE 10-2 Elemental Concentrations (ppm) in Horizons of Some Zonal Soils of the Soviet Union

Soil	Horizon	Depth (cm)	Co	Cr	Cu	Ni	Zn
Gleyed tundra	A_0	0–15	4.9	40	23	26	76
	A_1	15–25	1.5	—	11	10	—
	B	25–50	2.2	5	13	12	74
Podzol	A_1	0–18	14	180	8.3	24	33
	A_2	18–34	9.3	65	5.5	22	25
	B_1	34–60	6.6	290	5.4	16	25
	B_2	60–110	8.3	220	6.8	22	34
Grey forest	A_1	0–5	8.1	570	12	27	—
	A_2	20–25	6.4	140	10	17	28
	B_1	40–45	6.7	31	9.4	22	26
	B_2	75–80	7.8	440	8.4	2.2	30
	C	100–105	7.2	27	7.6	21	47
Chernozem	A_1	0.5	6.7	400	17	39	90
	A_2	24–32	5.6	630	12	33	90
	B	80–82	7.5	220	19	35	63
	C	128–144	12	200	16	48	—
Grey desert	A	0.5	3.4	570	5.1	10	39
	B	65–70	6.1	570	8.4	28	41
	C	160–170	8.8	350	12	29	—

SOURCE: Malyuga (1964).

TABLE 10-3   Horizons of Russian Soils with Maximum Enrichment of Trace Elements

Element	Tundra	Podzols	Grey forest	Chernozems	Chestnut	Serozems	Red
As	A, A_1	A, A_1	A_0, A_1	A	B	B	B
B	U	A_1	A_0, A_2	A	A	A	A
Ba	A_2, B	A	C	U	A	A	U
Cd	U	U	U	Y	Y	Y	Y
Co	U	B	B	U	U	U	B
Cr	—	B	B	A	A_1	A, B	A
Cu	U	C	A_0, A_1	A	A	A	U
Li	—	B	B	A, B	A, B	—	—
Mn	A_0, A_1	A_1	A_1	U	U	A_1	U
Mo	A	A	A	A	A	—	—
Ni	—	B	B	U	U	U	B
Pb	—	—	A_1	A_1	—	—	—
Rare earths	A_1	A_1	A_1	A_1	—	—	A_1
Sr	A_2, B	A	C	?	A	A	—
V	U	B	B	A	U	U	U
U	A	A	A	A	A	—	—
Zn	U	A_0, B	U	U	U	U	U

U = Uniform distribution.

SOURCE: Compiled from Vinogradov (1959).

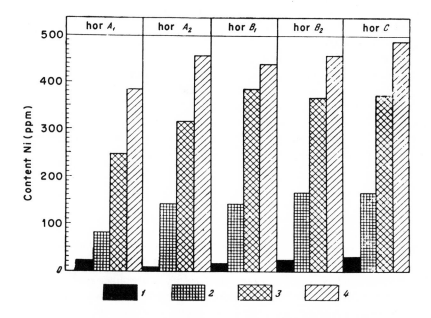

FIGURE 10-2   Distribution of nickel in various fractions of different horizons in brown forest soil. 1, 2.00–0.05 mm; 2, 0.05–0.02 mm; 3, 0.02–0.002 mm; 4, less than 0.002 mm.   From Malyuga (1964) and Painter et al. (1953). By courtesy of the Plenum Publishing Corp.

10-2   THE MOBILIZATION OF ELEMENTS IN THE PRIMARY ENVIRONMENT

Hypogene mobility occurs at depth under conditions of high temperature and pressure. The major constituents of the earth's crust form a sequence of minerals dependent on the prevailing temperature and pressure (Bowen, 1928). The minor elements usually occupy spaces in the lattices of these minerals according to the rules of *diadochic substitution*.

The residual fluids deposited as pegmatites or hydrothermal veins are usually extremely rich in trace elements. The same sequence of mobility is found in metamorphism when the last-formed minerals are the first to become liquid again as the temperature rises. The mobilization and transport of the elements in the primary environment is known as primary dispersion. The reader is referred to Hawkes and Webb (1962) for fuller discussion of this subject.

10-3   THE MOBILIZATION OF ELEMENTS IN THE SECONDARY ENVIRONMENT

Supergene mobility in secondary dispersion within the surface environment is of great importance in elemental differentiation within soils and takes place under conditions of low temperature and pressure. Mobilization is strongly influenced by Eh, pH, and the stability of the minerals which have to be decomposed. The factors which ultimately cause the breakdown of the minerals are mechanical, physical, chemical, and biological. Garrels (1960) has calculated the theoretical mobilities of trace elements in drainage waters from consideration of the physical chemistry of the associated mineral and ionic species. Equilibrium between mobile and immobile phases in the natural environment is, however, seldom encountered due to interferences from a wide variety of factors such as absorption onto clay minerals and humic material.

Perel'man (1967) has calculated a factor known as the *coefficient of aqueous migration* (K) which is a measure of the mobility of a particular element and is defined by the following expression:

$$K = \frac{M \cdot 100}{a \cdot N}$$

where M is the concentration of the element in drainage waters (mg/liter), N is the concentration of the element in rocks (%), and a is the mineral residue (%) contained in the water. This treatment takes into account the Eh and pH of the water. Table 10-4 shows the relative mobilities of the elements in the supergene environment. This table shows the pH dependence of the mobility of some elements, particularly the group cobalt, copper, mercury, nickel, silver, and uranium. The group molybdenum, selenium, uranium, and vanadium mobilizes readily under oxidizing conditions, since in all cases the higher oxidation states of these elements are much more mobile. This is particularly true for the UO_2^{2+} and U^{6+} ions.

TABLE 10-4   Relative Mobilities of the Elements in the Supergene Environment

Relative mobilities	Environmental conditions			
	Oxidizing	Acid	Neutral to alkaline	Reducing
Very high	Cl, I, Br, S, B	Cl, I, Br, S, B	Cl, I, Br, S, B, Mo, V U, Se, Re	Cl, I, Br
High	Mo, V, U, Se, Re, Ca, Na, Mg, F, Sr, Ra, Zn	Mo, V, U, Se, Re, Ca, Na, Mg, F, Sr, Ra, Zn, Cu, Co, Ni, Hg, Ag, Au	Ca, Na, Mg, F, Sr, Ra	Ca, Na, Mg, F, Sr, Ra
Medium	Cu, Co, Ni, Hg, Ag, Au, As, Cd	As, Cd	As, Cd	
Low	Si, P, K, Pb, Rb, Ba, Be, Bi, Sb, Ge, Cs, Tl Li	Si, P, K, Pb, Li, Rb, Ba, Be, Bi, Sb, Ge, Cs, Tl, Fe, Mn	Si, P, K, Pb, Li, Rb, Ba, Be, Bi, Sb, Ge, Cs, Tl, Fe, Mn	Si, P, K, Fe, Mn
Very low to immobile	Fe, Mn, Al, Ti, Sn, Te, W, Nb, Ta, Pt, Cr, Zr, Th, Rare earths	Al, Ti, Sn, Te, W, Nb, Ta, Pt, Cr, Zr, Th, Rare earths	Al, Ti, Sn, Te, W, Nb, Ta, Pt, Cr, Zr, Th, Rare earths, Zn, Cu, Co, Ni, Hg, Ag, Au	Al, Ti, Sn, Te, W, Nb, Ta, Pt, Cr, Zr, Th, Rare earths S, B, Mo, V, U, Se, Re, Zn, Co, Cu, Ni, Hg, Ag, Au, As, Cd, Pb, Li, Rb, Ba, Be, Bi, Sb, Ge, Cs, Tl

SOURCE: Andrews-Jones (1968).

10-4   FACTORS CONTROLLING MOBILIZATION WITHIN THE SOIL PROFILE
The mobilization and distribution of the elements within the soil profile is controlled by the following main factors:

1. Mobilization due to breakdown of soil minerals by various weathering agents and leaching through the profile
2. Adsorption of ions on to clay minerals and humus
3. Surface enrichment of elements by plant material (the biogeochemical cycle
4. Mobilization or fixation by soil microorganisms

Each of these factors will be considered in turn.

MOBILIZATION DUE TO BREAKDOWN OF SOIL MINERALS   The relative
ease of breakdown of soil minerals under chemical weathering is shown in Table

10-5.   The stability of the ions so formed is heavily dependent on the *ionic potential* (Z/r), where Z is the ionic charge and r is the ionic radius.   Ions with low chemical potentials (less than 2) tend to remain in solution whereas those with higher potentials precipitate more easily.   When the potential exceeds 12, oxyacid anions, such as phosphates and sulfates, tend to be formed.   The ionic potentials of the elements are shown in Figure 10-3.

The biosphere is a very important agency in the weathering of minerals and there is a wide range of organisms which are capable of producing this effect.

Pauli (1968) has summarized some of the main mechanisms of biologic weathering and has shown that the most powerful effects are produced by lower organisms such as lichens and bacteria.   Many plants increase the solubility of minerals by production of carbon dioxide and organic acids at their roots; others such as lichens are said to conduct a more mechanical attack by the production of mucous substances.   *Rhizocarpon geographicum* (Malyuga, 1964) attacks the garnet almandine with the formation of amorphous iron compounds and must be able to exert very powerful forces since garnets require attack with strong mineral acids before they can be dissolved.   Destruction of minerals by exudation of mucous substances is also achieved by the algae *Navicula miniscula* and *Nitschia palla*.

Bacteria have perhaps the greatest influence on the breakdown of minerals. The effect of nitrifiers and of *Bacillus megaterium* on calcium carbonate, magnesium carbonate, and phosphates is very well known.   Anaerobes which produce lactic and butyric acids are among others that attack calcite, magnesite, aragonite, dolomite, and siderite.   *Bacillus amylobacter* and *Bacillus extroguens* are strong corrosive agents for felspars with the consequent solubilization of potassium and silicon.

TABLE 10-5   Chemical Stability of Minerals in the Zone of Weathering

Degree of stability	Gangue minerals	Ore minerals
Very stable	Corundum, quartz, spinels, topaz, tourmaline, zircon	Chromite, gold, platinum, rutile
Stable	Alkali feldspar, andalusite, garnets, kyanite, muscovite, sillimanite, sodic plagioclase	Barite, cassiterite, galena, ilmenite, magnetite, monazite, tantalite
Fairly stable	Amphiboles, apatite, chloritoid, epidotes, pyroxenes, sphene, staurolite	Hematite, scheelite, wolframite
Unstable	Alkalic amphiboles, augites, biotite, calcic plagioclase, calcite, chlorite, dolomite, glauconite, gypsum, halite, hornblendes, olivine	Arsenopyrite, chalcopyrite, fluorite, molybdenite, pyrite, pyrrhotite-pentlandite, sphalerite

SOURCE: After Andrews-Jones (1968).

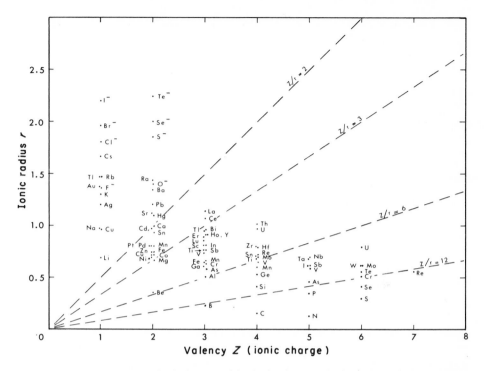

FIGURE 10-3   The ionic potentials (ionic charge ÷ ionic radius) of the elements. From Andrews-Jones (1968) based on data from Krauskopf (1967). By courtesy of the Colorado School of Mines.

ADSORPTION OF IONS ON TO CLAY MINERALS AND HUMUS   In typical cool-climate soils, the major proportion of the elements is adsorbed on to clay minerals or humus both of which have a high *cation-exchange capacity*.

The ion-exchange capacity of clays arises from their structures which usually are platelike in form and contain an excess of oxygen so that the overall charge on the crystal is negative.   Charged clay particles are known as *micelles* and attract positive ions by electrostatic forces.

The excess oxygen arises from the presence of hydroxyl groups attached to aluminum and which are exposed at the broken edges of most silicate clays.   The hydrogen of these hydroxyl groups dissociates slightly leaving a negative charge on the oxygen. Another mechanism for the production of this overall negative charge is substitution of an atom such as magnesium for aluminum within the lattice. Since an atom with a valency of 2 is substituting for one with a valency of 3, there is an unsatisfied valence of 1 unit and a consequent negative charge.

Reducing the size of the clay particles tends to increase the exchange capacity of the material. Table 10-6 shows the exchange capacity of a number of clay minerals and includes data for humus and various soils.

TABLE 10-6   Cation-Exchange Capacity of Clays,
              Humus, and Soils

Materials	Cation-exchange capacity (meq/100 g)
Kaolinite	3–15
Halloysite	5–50
Montmorillonite	80–150
Illite	10–40
Chlorite	10–40
Vermiculite	100–150
Organic fraction of soils	150–500
Podzolic soils (U.S.A.)	5–25
Chernozem (U.S.S.R.)	30–60
Black cotton soil (India)	50–80
Latosol (N. Rhodesia)	2–10
Gley soil (N. Rhodesia)	15–25

SOURCE: Hawkes and Webb (1962).

Anion exchange has so far not been discussed in considering clay minerals. Hydroxyl ions formed on the surface of clay particles are commonly available for exchange. The anion-exchange capacity of soils is apparently linked with the presence of iron and aluminum since removal of these elements quickly reduces adsorption of anions. Several elements such as boron, chromium, molybdenum, selenium, and vanadium often are present as soluble anions, and in this form may be adsorbed by clays.

Organic matter in soils also has an ability to concentrate cations. The effect of bonding will be either precipitation or mobilization depending on the solubility of the donor group. The exchange capacity of humus also increases with the pH of the environment. The structure of humic acids is not well known but formulae such as the following have been suggested (Malyuga, 1964): $C_{60}H_{35}(COOH)_{4-6}(OH)_{3-5}$ $(OCH_3)_{1-2}(CH{=}COH)$. At higher pH values, the ionization of the weak acid groups such as the carboxyl radical is increased and bonding can take place more readily. It is not certain whether bonding is of an ionic nature as with clay minerals or whether chelate links are involved with the carbonyl oxygen atom as the donor.

The exchange capacity of humus is far in excess of that of clay minerals. Whereas the capacity of clays seldom exceeds 100 meq/100 g, values of up to 500 meq/100 g can be obtained for humic material. In general, the presence of 1% humus in a soil indicates an exchange capacity of about 2 meq/100 g whereas the corresponding contribution from the same amount of clay would be only 0.1–1.0 meq/100 g.

THE BIOGEOCHEMICAL CYCLE   The accumulation of elements in the upper humic layer of soils is largely a function of the vegetation cover and is an immensely important factor in biogeochemical prospecting. In essence, the mechanism involves the uptake of elements by the root system, their passage through the aerial parts of the plant, and final deposition of the metal complexes in leaves or wood. As the leaves fall and wither, the more soluble components such as carbonates of the alkali metals and alkaline earths, sulfates, phosphates, and humic complexes of iron and manganese are leached by rain water and recycle through the soil profile. Sparingly soluble or insoluble compounds and complexes of other metals such as hydroxides and protein or humic complexes are retained in the humic horizons of the soil. The cumulative effect of this process is to produce a very significant enrichment of certain elements in the topmost layer of the soil.

Humic layers become particularly enriched in elements such as arsenic, beryllium, cadmium, cobalt, germanium, gold, lead, manganese, nickel, scandium, silver, tin, uranium, and zinc. The biogenic enrichment of such elements is known as the "Goldschmidt enrichment principle" (Goldschmidt, 1937) and has been deduced from the study of coal ash, which is presumed to be analogous with humus.

The above sequence of events is known as the *biogeochemical cycle* and is illustrated in Fig. 10-4.

Table 10-7 shows the relative enrichment of a number of elements in the humus layer of a Russian chernozem. It is noteworthy that the degree of enrichment follows the rules suggested by Irving and Williams (1953) and Basolo and Pearson (1958).

These rules can be summarized briefly as follows. The stability of metal complexes with organic matter will be largely independent of the ligand. For divalent cations, the stability of complexes is:

$$Pt > Pd > Hg > UO_2 > Be > Cu > Ni > Co > Pb > Zn >$$
$$Cd > Fe > Mn > Ca > Sr > Ba$$

For monovalent cations, the sequence is: $Ag > Tl > Li > Na > K > Pb > Cs$. The corresponding values for trivalent cations are: $Fe > Ga > Al > Sc > In > Y > Pr > Ce > La$.

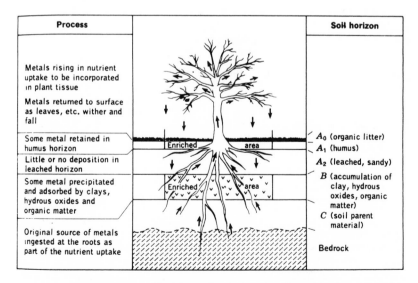

Process		Soil horizon
Metals rising in nutrient uptake to be incorporated in plant tissue		
Metals returned to surface as leaves, etc. wither and fall		
Some metal retained in humus horizon	Enriched area	A_0 (organic litter) / A_1 (humus)
Little or no deposition in leached horizon		A_2 (leached, sandy)
Some metal precipitated and adsorbed by clays, hydrous oxides and organic matter	Enriched area	B (accumulation of clay, hydrous oxides, organic matter) / C (soil parent material)
Original source of metals ingested at the roots as part of the nutrient uptake		Bedrock

FIGURE 10-4   The biogeochemical cycle. From Hawkes and Webb (1962).

TABLE 10-7 Relative Accumulation of Elements in the Humus Layer of a Russian Chernozem[a]

Horizon	Cr	V	Ni	Co	Zn
A (0–5 cm)	300	170	77	13	63
C (130–140 cm)	160	89	45	10	55
Enrichment (A/C)	1.9	1.9	1.7	1.3	1.1

[a] Concentrations in ppm.

SOURCE: Compiled from Vinogradov (1959).

The degree of enrichment of elements in the humic layer of soils depends very largely on the type of vegetation, since plants vary greatly in their specificity for accumulation of various elements. When several species of plants grow over soils with the same concentration of a particular element, the amounts of this element in different species can differ by several orders of magnitude. This is shown in Table 10-8 which gives data for nickel and selenium for various plant species.

MOBILIZATION AND FIXATION OF ELEMENTS BY MICRO-ORGANISMS
The major proportion of the mobilization and fixation of elements by living creatures is carried out by the soil microorganisms, particularly bacteria. Although bacteria form an insignificant proportion by weight of the total soil mass, their metabolic processes (mainly oxidative) can effectively handle large quantities of material.

The metabolism of all microorganisms is accelerated under conditions of high

TABLE 10-8 Variable Accumulation of Nickel and Selenium by Plants Growing in Soils Containing Constant Amounts of Each Element

Element	Species	Concentration in plant ash (ppm)	Concentration in soil (ppm)
Nickel	Cassinia vauvilliersii	2200	2500
	Myrsine divaricata	850	2500
	Gentiana corymbifera	780	2500
	Leptospermum scoparium	650	2500
	Stellaria roughii	350	2500
	Hebe odora	90	2500
Selenium	Astragalus pectinatus	4000	2
	Stanleya pinnata	330	2
	Aplopappus fremontii	320	2
	Gutierrezia sarothrae	70	2
	Zea mays	10	2
	Xanthium sp.	6	2
	Salsola pestifer	5	2
	Munroa squarrosa	4	2
	Helianthus annuus	2	2
	Malvastrum coccineum	1	2

SOURCES: Data for nickel from Lyon et al. (1970). Data for selenium from Williams (1938).

temperature and humidity. In well-aerated soils, most processes are oxidative, whereas these reactions are retarded in moist conditions.

Microorganisms play a major part in the cycles of carbon, nitrogen, phosphorus, and sulfur. Under reducing conditions, sulfides are formed and precipitate elements such as copper, iron, molybdenum, and zinc. The solubility of iron and manganese is particularly affected by soil organisms. Ferrous iron is oxidized to the ferric state under favorable conditions. In contrast, reducing bacteria produce the opposite effect and give the profile the characteristic grey streaky appearance of *gleyed* material. Manganese salts are similarly oxidized to insoluble manganese dioxide. Oxidation or reduction of other elements such as arsenic, molybdenum, selenium, tellurium, uranium, and vanadium can also be effected by microorganisms. For a fuller account of the role of microorganisms in biogeochemistry, the reader is referred to Zajic (1969).

Another method whereby elemental concentrations can be affected by bacteria is by uptake of the element by the organism itself so that there is residual depletion in other parts of the soil. Loutit et al. (1967) have shown that under favorable conditions the concentration of molybdenum in microorganisms from New Zealand soils is over 180 times that of the soil solution.

SELECTED REFERENCES

Primary and secondary dispersion: Hawkes and Webb (1962)
Biologic weathering: Pauli (1968)
Fixation or mobilization of elements by microorganisms: Zajic (1969)

CHAPTER ELEVEN
ELEMENTAL
UPTAKE BY
PLANTS

The reproducible uptake of soil elements by plants of the same species is an essential prerequisite for successful biogeochemical prospecting. The purpose of this chapter is to examine the mechanisms concerned with elemental accumulation by plants and to discuss the main factors that can influence this process.

11-1   ESSENTIAL AND NONESSENTIAL ELEMENTS

The presence of an element in plant tissues does not necessarily imply that it exercises a useful function in the species concerned. It is probably true to say that many more nonessential elements are taken up by plants than essential nutrients although some elements now thought to be nonessential may later prove to have a role in plant nutrition.

By the late nineteenth century, it had been established that carbon, hydrogen, oxygen, nitrogen, phosphorus, sulfur, potassium, calcium, magnesium, and iron were universally essential for plant growth. Later, another six elements: boron, chlorine, copper, manganese, molybdenum, and zinc were added to the list.

It is probable that in time other elements will also prove to be essential. These other elements have not yet been discovered, probably because the plant requirements may be so low that it has not yet been possible to prepare nutrient solutions which are sufficiently purified from the element concerned in order to make experiments meaningful.

Apart from the *universal* nutrients listed above, there are other elements such as selenium, cobalt, silicon, and vanadium which apparently have a beneficial effect on some plants though they may not be essential elements.

The role of the essential elements in plant nutrition is summarized in Table 11-1. Boron is not included in this listing because its role in plant nutrition is not fully understood. Carbon, hydrogen, and oxygen are also not mentioned since their function as constituents of organic matter is so obvious.

TABLE 11-1   Physiological Roles of Essential Elements in Plants

Element	Physiological role
Calcium	Component of calcium pectate in cell walls and required for cell membrane stability
Chlorine	Electrolyte function
Copper	Enzyme activator
Iron	Found in prosthetic group of respiratory enzymes
Magnesium	Metallic constituent of chlorophyll
Manganese	Enzyme activator
Molybdenum	Found in metalloflavin enzymes
Nitrogen	Component of proteins
Phosphorus	Found in nucleic acids, phospholipids, and coenzymes
Potassium	Osmotic regulator in cell vacuoles
Sulfur	Found in proteins
Zinc	Found in metalloenzymes

11-2   ABSORPTION OF IONS BY PLANTS

There are three main mechanisms whereby ions may be absorbed by plants. Two of these pathways involve uptake at the root system; the third, foliar absorption, involves the aerial parts of the plant.

Uptake at the root system involves either diffusion into the plant from the soil solution or cation exchange at the surface of clay minerals. Cation exchange is the more important of the two processes and takes place by the production of carbon dioxide as a result of respiration processes. The carbon dioxide reacts with water and liberates hydrogen ions which exchange with cations held upon the clay mineral. When the cations reach the root tip, they are again replaced by hydrogen ions and the cycle is repeated (Keller and Frederickson, 1952).

Some ion absorption also occurs by simple diffusion into the cells of the root tips. A surprising feature of the accumulation of ions by roots is that their concentration in the cell fluid is often many times greater than the soil solution. This is known to be a metabolically mediated process requiring the expenditure of cellular energy.

Most plants are selective in their accumulation of trace elements. Toxic metals such as lead or uranium can be precipitated as sulfate or torbernite (copper-uranyl phosphate), respectively.

The ability of plant species to restrict uptake of a toxic element is known as an *exclusion mechanism*. In such cases, the amount of the element in the aerial parts of the plant (twigs and leaves), remains at a constant level whatever the amount of the toxic element in the soil. This exclusion mechanism ultimately breaks down at a certain threshold concentration of the element in the soil. Above this threshold, increased amounts of the element are accumulated over a short concentration range until the amount in the soil is completely toxic to the plant. This is illustrated in Figure 11-1 which shows the effect on *Triodia pungens* of increasing amounts of copper, lead, and zinc in the soil. The exclusion mechanism for lead is very marked, whereas that for copper is obviously only partial. The plant will apparently take up zinc in unlimited quantities without evidence of exclusion. Unlimited uptake of any element is, however, unlikely. In Figure 11-1, the values of zinc in the soil were obviously not high enough to reach the threshold level.

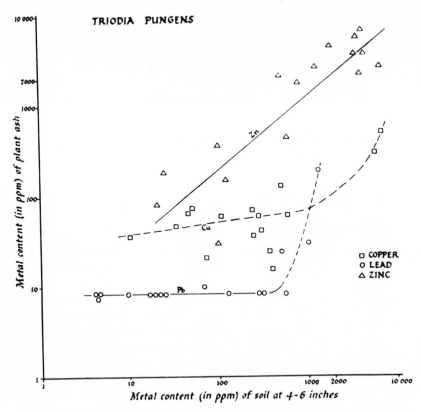

FIGURE 11-1 Relationship between the copper, lead, and zinc contents of plant ash of *Triodia pungens* and associated soils. From Nicolls et al. (1965). By courtesy of M. M. Cole, J. S. Tooms, and the Institution of Mining and Metallurgy, London.

Plants which operate a total exclusion mechanism are obviously useless for biogeochemical prospecting. But those that have partial exclusion mechanisms or none at all are often quite suitable for this purpose. When a partial exclusion mechanism is involved, the result is an all-or-nothing indication of the outline of the anomaly in which only the values corresponding to high levels in the soil will be recorded.

11-3 TRANSLOCATION OF ELEMENTS WITHIN THE PLANT
The ions absorbed at the roots of the plant are usually translocated in an upward direction toward the leaves. The work of Stout and Hoagland (1939) seems to indicate that the *xylem* is the chief agency for this transport.

There is evidence for a reverse flow of some nutrients from the leaves back to the stems or younger leaves, usually just before exfoliation. Movement of minerals from the leaves occurs via the *phloem*. There is also some evidence for periodic daily circulation of some minerals.

The form in which mineral salts are translocated in plants is extremely variable. Work by Lyon et al. (1969) has shown that the uptake of chromium by the New Zealand species *Leptospermum scoparium* takes place as the oxalato complex. Whitehead et al. (1971) have shown that uranium in *Coprosma australis* is bound mainly as a complex with protein.

An extremely important factor in plant nutrition is the *availability* of the elements present in the soil. For some elements, very little of the total content in the soil can be utilized by the plant. This can be due to a large number of factors including drainage, pH, Eh, the nature or absence of clay minerals, antagonistic effects of other ions, and presence of complexing agents in the soil.

A number of different tests have been devised for determining the availability of elements in soils. These procedures involve shaking the soil with various solutions such as ammonium acetate or acetic acid and measuring the concentration of the elements in the aqueous phase after equilibration. In practice, the only reliable way of measuring availability is by pot trials with suitable test plants and subsequent analysis of the elements in the plant material.

11-4 BIOGEOCHEMICAL PROVINCES

Variations in the health, form, and vitality of plants and animals under the influence of various chemical ecological factors in the soils, rocks, and waters from which they obtain their nutrients have been known for many years. An excess or deficiency of a particular element or group of elements is usually linked with some form of endemic disease or condition which readily distinguishes the district from others in the neighborhood. Classical examples of this are the effects of iodine deficiencies on humans and animals and the effects of serpentine soils on vegetation. Vinogradov (1964) has assigned the name *biogeochemical provinces* to such areas. The following quotation from some of his earlier work is adapted from Malyuga (1964).

> We apply the name biogeochemical provinces to regions on the earth which differ from adjacent regions with respect to their content of chemical elements, and which as a result experience a different biological reaction by local flora and fauna. In extreme cases, as a result of a marked deficiency or excess of the content of any chemical element (or elements) in the province, plants and animals will experience biogeochemical endemias.
>
> . . .
>
> By regions of the earth, we mean rocks, soils, and water basins which are populated by organisms and which have limits in time and in space. We include in this definition parts of the sea, atmosphere, or sea and atmosphere together. With respect to the content of one or more chemical elements in a particular region, that is, in rocks, soils, water and the organisms of this region, we mean either the usual normal level or an inadequate or excess content of a particular chemical element or several elements (or their compounds). It should be noted that it is of importance to know not only the absolute content of an element in a particular region, in its rocks and soils,

but also the content of any particular form of this element, for example Fe^{2+} or Fe^{3+}; Mn^{2+} or Mn^{4+}; Se^{2+} or Se^{4+}. Thus we often encounter not only an absolute shortage or excess, but also a relative deficiency or excess, depending on the degree of availability of any particular chemical element to particular types of flora and fauna. For instance, cobalt and many other chemical elements in areas where there is an alkaline soil reaction, are quite unavailable to plants. On the other hand, under these conditions, molybdenum is easily taken up as a result of easily soluble molybdates, while in an acidic medium this element is less available to plants due to its leaching into compounds which are not easily soluble.

The biological reaction of flora and fauna of a particular region arising under the influence of an excess or deficiency of any chemical element (or elements) is the most important and basic criterion for the delimitation of a biogeochemical province.

Biogeochemical provinces are said to be *zonal* (strongly influenced by climate and soil type) or *intrazonal*. Zonal provinces usually involve deficiencies in elements such as calcium, cobalt, and iodine, due to the nature of the zonal soils in the area. Intrazonal provinces are often influenced by local enrichment of metals due to the existence of ore bodies and their associated dispersion halos.

An intrazonal biogeochemical province, as defined by Vinogradov, represents an obvious target area for geobotanical or biogeochemical prospecting procedures since the region represents a zone where there will be variations and mutations in the flora so that either or both techniques may be applied. It may well be that the only obvious changes are those involving increased levels of certain elements in the vegetation, but this is the very factor which justifies the use of biogeochemical prospecting in the province.

11-5   ACCUMULATION OF ELEMENTS BY PLANTS

Nearly all the chemical elements have been determined in vegetation and it is true to say that for every element there will be at least one species capable of concentrating it in a spectacular manner. The volume of literature on elemental uptake by plants is now so vast that the underlying principles are liable to be confused by the great abundance of the data. When the information is sifted and the more reliable data are examined, certain patterns become apparent.

The preferential enrichment of certain elements by plants, known as the *Goldschmidt enrichment principle* (Goldschmidt, 1937), is illustrated in Table 11-2. The table shows the mean concentrations of a number of elements in plants and soils. The ratio of these two quantities is a measure of the *biogenic* characteristics of each element. To say that an element is biogenic implies either that it has a physiological role in the species concerned or else that it is concentrated to an appreciable degree whether or not it takes part in the plant's metabolism. Elements which are accumulated without requirement are known as *ballast elements* (*Ballaststoffe*) (Frey-Wyssling, 1935). In Table 11-2 it is assumed, somewhat arbitrarily, that elements with an enrichment coefficient of 0.10 or greater are biogenic and that those with coefficients less than 0.01 are nonbiogenic. Between these two extremes lies a third group with intermediate properties.

TABLE 11-2   Mean Elemental Concentrations in Plants and Igneous Rocks

Biogenic elements (ppm)				Intermediate elements (ppm)				Nonbiogenic elements (ppm)			
Element[a]	Plant[b]	Rock	Ratio	Element	Plant	Rock	Ratio	Element	Plant	Rock	Ratio
B*	5	3	1.70	Mo*	0.65	15	0.040	Na	200	28,300	0.007
S*	500	520	0.96	Mg*	700	21,000	0.034	Rb	2	310	0.007
Zn*	70	80	0.90	Ni	3	100	0.030	R.E.[c]	0.40	148	0.003
P*	700	800	0.88	Co	0.40	18	0.020	Cr	0.40	120	0.003
Mn*	400	1,000	0.40	U	0.05	2.6	0.020	Li	0.10	65	0.0015
Ag	0.05	0.2	0.25	Fe*	500	40,000	0.012	Si	1500	271,000	0.0006
Ca*	5,000	36,000	0.14					V	1	150	0.0006
Sr	20	150	0.13					Ti	2	6,400	0.0003
Cu*	9	70	0.13					Al	20	81,000	0.0003
K*	3,000	26,000	0.12								
Ba	30	250	0.12								
Se	0.1	1	0.10								

[a]Essential elements are asterisked.   [b]Dry-weight basis.   [c]Rare earths.

SOURCES: After Hutchinson (1943) and Cannon (1960b).

Hutchinson (1943) has shown that the biogenic elements occupy certain restricted areas on an ionic potential diagram. The data are shown in Figure 11-2 from which it will be noted that elements of low ionic potential, which form soluble cations, and those of high potential, which form soluble anions, correspond well with those elements of Table 11-2 that are enriched in vegetation to an appreciable degree.

Although the credit for the enrichment principle is usually ascribed to Goldschmidt, most Russian scientists claim that the originator was Vernadsky (1934). There can be no doubt that Vernadsky discussed the biogenic migration of the chemical elements well before Goldschmidt, who formulated the principles in a more precise form at a later date (Goldschmidt, 1937). Other workers (Shaw, 1960; Samoilova, 1961; Kovalevsky, 1962a, 1963; Glazovskaya, 1964; Poluzerov, 1965; Boichenko et al., 1968) have discussed the principles of biogenic migration further and some have formulated new theories, but the fundamental validity of the Goldschmidt principle still remains.

11-6   THE EXTENT OF ELEMENTAL ACCUMULATION BY PLANTS AND ITS SIGNIFICANCE FOR BIOGEOCHEMICAL PROSPECTING

It has already been stated that some plants can concentrate elements of the periodic table to a spectacular degree. Over ore deposits this enrichment can be even greater so that in many cases the ash of the plant contains more of the element than the soil itself. The extent of possible enrichment of elements is documented in Table 11-3.

An evaluation of the enrichment coefficients of various plants gives some indication of the likely behavior of the species when growing over mineralized ground and indeed data are already available for normal background levels of various elements in vegetation, particularly in North America (Cannon, 1960b; Warren, 1962). It is not necessarily an advantage that enrichment coefficients should be high, because this only increases the background level in plants and may render an anomaly more

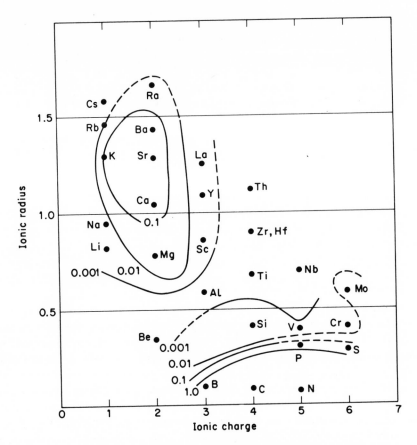

FIGURE 11-2 Enrichment coefficients of terrestrial plants for various elements in the earth's crust in relation to their ionic potential. From Hutchinson (1943). By courtesy of G. E. Hutchinson and the American Society of Naturalists.

TABLE 11-3 High Concentrations of Elements in the Ash of Plants Growing over Ore Bodies

Element	Plant	Ash content (ppm)	Reference
Cobalt	*Crotalaria cobalticola*	17,700	Duvigneaud (1959)
Copper	*Becium homblei*	2,500	Horizon (1959)
Chromium	*Pimelea suteri*	26,500	Lyon et al. (1968)
Lead	*Beilschmiedia tawa*	7,300	Nicolas and Brooks (1969)
Molybdenum	*Olearia rani*	1,600	Lyon and Brooks (1969)
Nickel	*Alyssum bertolonii*	100,000	Minguzzi and Vergnano (1948)
Selenium	*Astragalus pattersoni*	46,000	Cannon (1960a)
Uranium	*Uncinia leptostachya*	25,000	Whitehead and Brooks (1969b)
Vanadium	*Astragalus preussi*	1,680	Cannon (1960a)
Zinc	*Viola calaminaria*	15,000	Linstow (1929)

difficult to detect. This has been demonstrated by Polikarpochkina et al. (1965) who showed that better results were obtained with weakly absorbed elements such as bismuth, gold, and molybdenum than with strongly absorbed elements such as copper and zinc.

As an example of very different behavior by two elements, uranium and zinc, may be considered. The normal background level of uranium in plants is about 1 ppm. Whitehead and Brooks (1969b) have reported uranium values ranging from 1.5 to 1000 ppm in the ash of the New Zealand shrub *Coprosma australis*. The response of this species to uranium is therefore very good and biogeochemical prospecting with this plant can probably be carried out with some success. Similar results have been obtained for *Olearia rani* (Brooks and Lyon, 1966; Lyon and Brooks, 1969), where molybdenum values ranged from 1 to 1600 ppm.

In the case of zinc, plants growing in unmineralized ground already have a high content of this element in the ash (about 1400 ppm) so that levels in the soil have to be extremely high before a significant increase in the plant content will become evident. The situation is complicated still further by the fact that biogeochemical prospecting for elements that are essential nutrients (such as zinc) often results in failure due to factors enumerated in Chapter 14.

When the enrichment coefficient for an element is particularly low, as in the case of niobium, tantalum, or zirconium, the reason for this is usually the extreme immobility of the ions. In such cases, it is unlikely that sufficient quantities of the elements will ever be accumulated by plants in order to render biogeochemical prospecting a feasible proposition.

Extreme enrichment of elements in plants does afford one or two possible advantages in future work. If the enrichment is sufficiently great, it should be possible to devise simple field tests (e.g., color reactions or gamma-source X-ray fluorescence) to test vegetation directly without bringing it back to the laboratory for analysis.

11-7   FACTORS AFFECTING ELEMENTAL ACCUMULATION BY PLANTS

It should be reemphasized that the main requirement of successful biogeochemical prospecting is that the plant species selected for the project should accumulate the element concerned in a *reproducible* manner and that the degree of accumulation should be in direct proportion to the concentration of the element in the soil. Such a situation is seldom found under natural conditions, although some plants can approach the ideal fairly closely. Provided that the contrast between background and anomalous concentrations is sufficiently great, it is probably true that *all* the vegetation in a given area will reflect mineralization to some extent. The selection of a suitable species for biogeochemical prospecting should, however, be far from haphazard and should be based on the results of a preliminary orientation survey (see Chapter 12) after taking into account some of the many factors that can affect elemental uptake by plants.

There are probably up to 20 different variables which can affect elemental accumulation by vegetation. There are, in fact, so many of these factors that it is indeed surprising that biogeochemical methods of prospecting have ever had success in the past. The field worker should be aware of all these variables, even though some of them may be minor in their effects. Use of suitable procedures will eliminate or at least reduce the effects of most of these variables; they are not, therefore, as serious a problem as might be expected. Some of these factors will now be considered.

TYPE OF PLANT SAMPLED   In biogeochemical prospecting the nature of the plant sampled is of great importance and should be studied carefully. There has already been some discussion on the selectivity and differences in uptake of elements by different species. Plants differ considerably not only in their relative power to accumulate elements from a soil of given composition but also in their response to increasing or decreasing amounts of the soil constituents under consideration.

Figure 11-3 shows the different response of three species to increasing amounts of molybdenum in the soil. Clearly, *Medicago hispida* would be the best choice for biogeochemical prospecting since, of the three species in the figure, it gives the greatest response. It is also clear that the three plants would not be interchangeable during a prospecting survey because their relative uptakes of molybdenum are very different.

Although the plots in Figure 11-3 are approximately linear over the range of molybdenum values studied, there is no guarantee that linearity would have been preserved at still higher molybdenum concentrations. There will ultimately be a point where the amount of the element in the plant will either remain constant or will tend toward "infinity" as the plant's regulatory mechanism breaks down. The result of a trend toward infinity (as shown in Figure 11-1 for lead) is that the plant will not grow at a certain threshold concentration of the element in the soil.

From the literature and from the author's own experience in New Zealand,

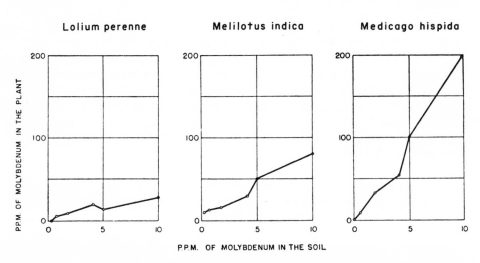

FIGURE 11-3   Relative uptake of molybdenum by three different plant species growing under the same conditions. From Carlisle and Cleveland (1958) based on data by Barshad (1951). By courtesy of the California Division of Mines and Geology.

TABLE 11-4   Enrichment Coefficients for Uptake of Radioactivity by Different Species of New Zealand Plants

Class	Species	Plant	Soil	Plant/soil
		Plant	Soil	Plant/soil
Bryophytes	Marchantia berteroana	1421	663	2.1
Ferns	Blechnum capense	939	231	4.1
	Dicksonia lanata	921	382	2.4
Monocotyledons	Cordyline banksii	95	87	1.1
	Uncinia leptostachya	3817	1732	2.2
Dicotyledons	Carpodetus serratus	149	585	0.2
	Coprosma arborea	24	3001	0.1

SOURCE: After Whitehead and Brooks (1969b).

elemental accumulation depends to some extent on the class of plant involved. In general, the smaller and more primitive the species, the greater the enrichment coefficient for a particular element. This is demonstrated in Table 11-4 which shows the enrichment coefficients of several plant species for the accumulation of uranium from a uraniferous area of New Zealand. One reason for the different degree of accumulation is probably that elemental concentrations in tissues may well be a function of the translocation distance. Hence, in the leaves of a large tree with a long translocation distance between the roots and outermost branches, elemental concentrations should be far lower than in the leaves of lowly bryophytes where the ions only have to travel a few millimeters.

PLANT ORGAN SAMPLED   Variation in elemental concentrations in different plant organs has been confirmed by the author's work in New Zealand. This work has corroborated the earlier findings of Warren and Delavault (1949) that leaves of plants usually have higher concentrations of trace elements. High concentrations of an element in a particular organ do not necessarily imply that this is the best part of the plant to sample for biogeochemical prospecting. It may well be that an organ with a lower content of the element sought gives a more reproducible correlation with the amount of this element in the soil.

Carlisle and Cleveland (1958) have shown that in North American vegetation the elemental contents of tree organs decrease in the sequence leaves, twigs, cones, wood, roots, and bark. Warren and Delavault (1949) have reported that (for copper and zinc at least) needles, cones, and leaves of trees give inconsistent data. The best results were obtained with twigs 1 to 3 years old and with diameters between $\frac{1}{8}$ inch to $\frac{1}{4}$ inch.

In fairly extensive studies on the above problem, using native New Zealand species, the author has found little difference in the reliability of leaves compared with twigs in prospecting for copper, molybdenum, nickel, uranium, and zinc. An exception was that twigs of *Beilschmiedia tawa* were more reliable than its leaves in prospecting for lead (Nicolas and Brooks, 1969). The lack of differentiation between leaves and twigs in New Zealand may be partly due to the fact that native trees are evergreen so that any sufficiently large sample of leaves will contain individuals of different ages. This factor tends to mitigate against seasonal effects such as are found in deciduous leaves. The equable nature of the New Zealand climate may also be a factor since in many parts of the country there is a 12-month growing season with the result that metabolism, and hence mineral uptake, will be more uniform throughout the year than in harsher climates.

Within individual organs there is sometimes strong evidence for antagonistic effects between different pairs of elements. Thus, care should be taken to sample organs in which these effects are minimal. Figure 11-4 shows the copper and lead contents of a series of tree rings from the New Zealand species, *Beilschmiedia tawa*. The different trends of the distributions of the two elements are very obvious although the reasons for them are not known.

AGE OF PLANT OR PLANT ORGAN   Of all the factors that influence the uptake of trace elements by plants, the age variable has received least attention. There is some evidence for appreciable variation in the elemental content of plants of the same species but of different ages. There is also strong evidence that the elemental content of certain plant organs, particularly leaves, is influenced by seasonal factors. A classical investigation into this problem was carried out by Robinson (1943), who studied the uptake of rare earths by the hickory tree by

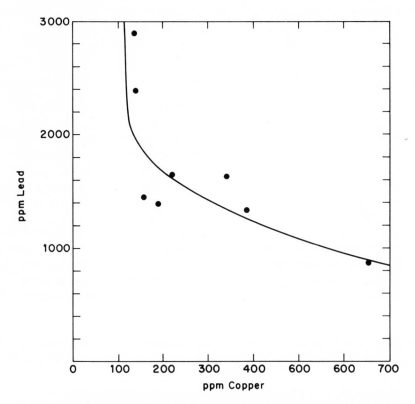

FIGURE 11-4   Example of mutual antagonism of copper and lead in the ash of trunk sections of *Beilschmiedia tawa*. Data from Nicolas and Brooks (1969).

checking the same specimen at different times of the year. He found that on June 1st, the content of the leaves was 174 ppm. A month later this had risen to 634 ppm; and on October 1st of the same year, the level had reached 981 ppm. Robinson and Edgington (1942) also observed a doubling of the boron content of hickory leaves between the spring and fall. Seasonal variations in leaves have also been noticed for molybdenum (Barshad, 1951) and for copper and zinc (Warren and Howatson, 1947).

Work in the Soviet Union has also revealed seasonal differences in the concentration of elements in leaves. Malyuga (1939) has reported that concentrations in the fall are two or three times greater than in the spring. Chebaevskaya (1960) has shown that the content of cobalt, copper, and nickel increases toward the time of development of the reproductive organs.

It now seems to be well established that the metal content of many *deciduous* leaves rises to a maximum just before exfoliation; this is probably a convenient mechanism whereby plants can get rid of unwanted ballast elements. One of the most convincing proofs of this seasonal variation has been furnished by Guha (1961). Figure 11-5 shows the variation of six elements in leaves of 18 deciduous species studied by Guha over a 12-month period. Increases in boron, iron, and manganese were noted over the growing season; levels of copper, molybdenum, and zinc decreased in many cases. Some of Guha's findings are in contradiction with those of other workers, but this may be a factor of species differences.

Most of the recorded literature on this subject has indicated a tendency for elements to be concentrated in actively growing tissues such as shoots and young leaves. Lowest levels are usually in the sapwood and heartwood of tree trunks.

HEALTH OF THE PLANT   The ability of a plant to accumulate trace elements is partly a function of its health. There is sometimes a tendency for the content to increase due to breakdown of the regulatory mechanisms of the species. Unfortunately, it is not always possible to predict whether disease will increase or decrease elemental accumulation. For this reason, unhealthy plants, although a useful guide in geobotany, should be avoided during biogeochemical prospecting operations.

THE pH OF THE SOIL   Although there are many variables which can adversely affect the reliability of the biogeochemical method, the pH factor predominates. Elemental uptake of elements by plants can be affected by the pH of the soil. How the availability of a number of elements is affected by pH is demonstrated in Figure 3-3. Drastic changes in pH can affect availability to a significant degree since elements can be either precipitated or solubilized depending on a number of soil factors. Also, pH can control availability by its effect on the exchange capacity of clay minerals. The influence of pH on the uptake by plants of elements of economic importance will now be discussed briefly.

Copper is readily immobilized in alkaline soils (Jamison, 1942). This worker found a 100% recovery of added copper at pH 4.0; at pH 6.0, recovery was only 33%. Warren, Delavault, and Irish (1952a) have reported a similar pH effect on the availability of copper to plants.

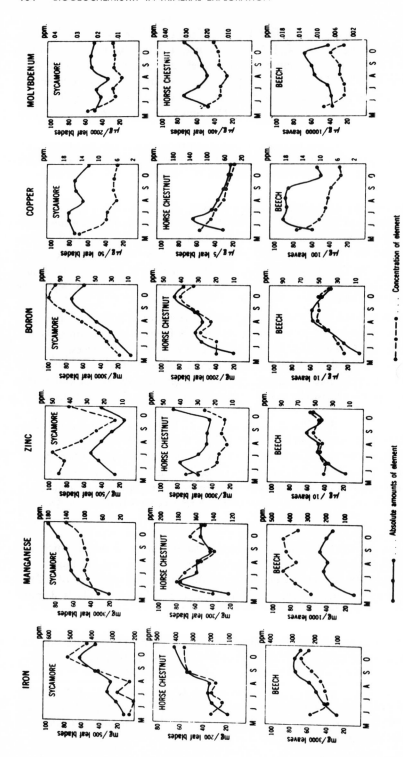

FIGURE 11-5   Seasonal variation of minor elements in the leaves of three species of deciduous trees. From Guha (1961). By courtesy of M. M. Guha.

Zinc behaves similarly to copper in accumulation by plants. Robinson et al. (1947) have reported that various trees depress their uptake of zinc in alkaline soils. Aldrich and Turrell (1951) reported increased availability of this element in acid soils.

Mulder and Gerretsen (1952) have shown that manganese is highly soluble below pH 5.5 and that, at higher pH values, the metal is converted to insoluble manganic oxides. Similar observations have been made by Aldrich and Turrell (1951), Brenchley (1936), and Fujimoto and Sherman (1948).

The availability of tungsten and molybdenum is increased at high pH ranges due to the formation of double tungstates and molybdates. These elements are precipitated in the pH range 3–5 (Clarke, 1924; Barshad, 1948; Hartsock and Pierce, 1952).

Table 11-5 shows the pH dependence of the precipitation of various ions as hydroxides. From this table it could be assumed that changes in pH would seriously invalidate biogeochemical prospecting data. But, provided that these changes are not too great over the survey area, the effect appears to be fairly minor. Experience in New Zealand has shown that even in prospecting for molybdenum (Brooks and Lyon, 1966; Lyon and Brooks, 1969), the availability of which is strongly influenced by pH, a range of 4.8 to 6.8 in the soils was not sufficient to affect the reliability of the method to any noticeable degree.

The pH dependence of elemental uptake by New Zealand plants has been investigated by Timperley et al. (1970a) under field conditions. The enrichment coefficients for copper, nickel, and zinc were obtained for about 300 samples of four species and were correlated statistically with the pH values of the soils. Clearly, a significant correlation would imply a pH dependence of metal uptake. Values were obtained for the correlation coefficient (r) between the two variables (see Chapter 13) and are shown in Table 11-6. In only one case was there an appreciable pH dependence of uptake. This was for nickel by *Weinmannia racemosa*. A value for r of -0.32 implied a highly significant inverse relationship. In other words, nickel accumulation decreased as the pH was raised.

It has already been mentioned that large changes in pH values of the soil could seriously invalidate the results of a biogeochemical survey. However, in a given area the pH of the soil depends to a large extent on relief and the nature of the bedrock. Provided that these do not change appreciably, little difficulty should be experienced in field work. Orientation work on a type of bedrock different from that in which the ore body is expected would, in any case, probably not be a

TABLE 11-5   pH Values at Which Ions Are Precipitated as Hydroxides

Ion	pH	Ion	pH	Ion	pH
Fe^{3+}	2.0	Cr^{3+}	5.3	Pb^{2+}	6.8
Sn^{2+}	2.0	Cu^{2+}	5.3	Zn^{2+}	5.3–6.8
Th^{4+}	3.5	Fe^{2+}	5.5	Hg^{2+}	7.3
Hg^{+}	3.0	Cd^{2+}	6.7	Ag^{+}	7.5–8.0
Al^{3+}	4.1	Ni^{2+}	6.7	Mn^{2+}	8.5–8.8
UO_2^{2+}	4.2	Co^{2+}	6.8	Mg^{2+}	10.5

SOURCE: Malyuga (1964).

TABLE 11-6   Correlation Coefficients (*r*) for the Relationship Between the pH of the Soil and the Enrichment Coefficients of Some New Zealand Trees for Various Elements

		Copper		Nickel		Zinc	
Species	Number	*r*	Significance	*r*	Significance	*r*	Significance
Nothofagus fusca	49	−0.12	NS	0.27	PS	−0.04	NS
Nothofagus menziesii	48	−0.02	NS	−0.25	PS	−0.11	NS
Quintinia acutifolia	93	−0.02	NS	−0.21	S	−0.04	NS
Weinmannia racemosa	103	−0.10	NS	−0.32	S*	−0.07	NS

NS = not significant
PS = possibly significant
S* = highly significant

SOURCE: Timperley et al. (1970a).

sound practice unless of course a contact between two formations was being examined.

DEPTH OF ROOT SYSTEM   Plants are the only parts of the prospecting prism (see Chapter 1) which extend through several of the layers simultaneously. It is claimed (Malyuga, 1964) that the main advantage of biogeochemical prospecting compared with other geochemical methods lies in its power of penetration through a nonmineralized overburden. This penetrating power in turn depends on the depth of the root system, the ability of the plant to translocate minerals from its roots to its aerial parts, and the extent of the dispersion halo which itself depends to some degree of fluctuations in the water table. The depth of root systems is largely a function of climate and, by inference, the type of soil involved.

There are many references to the effective depth of root systems in general. Weaver (1919) assessed the mean depth of forest shrubs and herbs as being not greater than 18 in.; Kelley (1923) reported an effective depth of not more than 8 in. for maximum absorption of elements by this type of plant. Warren and Howatson (1947) concluded that trees were no more effective than shrubs in penetrating a nonmineralized overburden. Data from Cannon (1960a, 1964) and Malyuga (1964) would tend to contradict Warren's findings.

Table 11-7 lists the root depths of various plant species and it will be seen that depths of up to 200 ft have been recorded. The latter figure was for a live root found in a mine working on the Colorado Plateau (Cannon, 1960a). The root depth of a plant is essentially a function not only of climate but also of whether the species is *phreatophytic* or *xerophytic*. Phreatophytes are plants which depend on the zone of saturation below the water table for their moisture; therefore, in desert areas they have extremely deep root systems. Xerophytes depend on surface water from rainfall and have shallow root systems. Clearly, phreatophytes will, in general, be more useful than xerophytes in prospecting for deep deposits. In some cases plants with extremely shallow root sytems can still indicate mineralization at depth. Cannon (1960a) has reported species of *Astragalus* which have geobotanically indicated uranium deposits at a depth of 60 to 70 ft. To some extent the suitability of phreatophytes also depends on the type of root. For example, a species with a long tap root may well be able to give evidence for mineralization

TABLE 11-7   Depth of Root Penetration of Woody Plants and Shrubs in Different Climatic Zones

Zone	Plant	Depth of roots (feet)
Tundra	*Betula nana*	0.8
	Larix sibirica	1.2
Forest podzol	*Picea excelsa*	6
	Pinus sylvestris	7.5
	Populus alba	9
Forest-steppe and steppe	*Pinus sylvestris*	11.4
	Betula verrucosa	12
Desert and semidesert	*Actinea acaulis*	0.8
	Artemisia spinescens	5
	Tamarix ramosissima	15
	Anabasis aphylla	15
	Quercus sp.	28
	Artemisia tridentata	30
	Haloxylon aphyllum	30
	Alhagi pseudalhagi	45
	Sarcobatus vermiculatus	57
	Atriplex canescens	62
	Pinus ponderosa	80
	Juniperus monosperma	200

SOURCES: Cannon (1964) and Malyuga (1964).

at depth but may fail to pick up an anomaly near the surface. This is illustrated in Figure 11-6 which shows the response of xerophytes and phreatophytes to ore bodies and dispersion halos at different depths.

Translocation of metals from the roots to the aerial parts of plants has been mentioned as a factor affecting the usefulness of a species for biogeochemical prospecting. In general, the longer the root system, the less the enrichment of trace elements in the upper part of the plant. This factor is illustrated in Table 11-8 which gives the uranium and vanadium contents of roots compared with aerial parts of various species. The difference in values between shallow and deep-rooted specimens is probably only a reflection of the greater difficulty of translocation of ions through a long root system.

In sampling vegetation, care should be taken to select leaves or twigs at several points around the circumference of the specimen studied. This is because root

TABLE 11-8   Uranium and Vanadium Contents of Roots Compared with Aerial Parts of Plants

Species	Type of root	Uranium in ash (ppm)			Vanadium in ash (ppm)		
		Roots	Tops	Ratio	Roots	Tops	Ratio
Juniperus monosperma	Deep	1600	7.8	200	3000	20	150
Juniperus monosperma	Deep	140	2.0	70	4000	50	80
Quercus gambeli	Deep	190	10.0	19	1700	90	19
Juniperus monosperma	Shallow	7	1.2	5.6	110	54	2
Atriplex confertifolia	Shallow	5	3.0	1.6	90	10	9
Astragalus preussi	Shallow	70	70	1.0	2600	3000	0.8

SOURCE: Cannon (1960a).

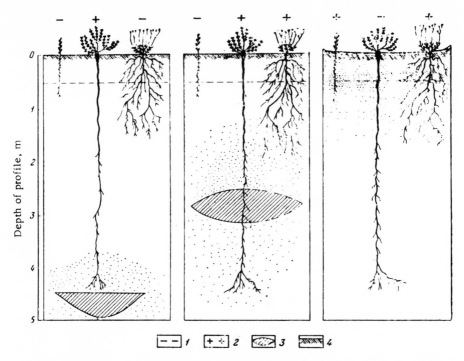

FIGURE 11-6   Plant indicators of boron as a function of the type of root system. 1, background content of boron in plants; 2, high boron content in plants; 3, horizons enriched in boron; 4, soil. From Malyuga (1964). By courtesy of the Plenum Publishing Corp.

systems tend to translocate ions to aerial parts situated on the same side of the plant as themselves. It could well be possible that roots on one side of the plant were in contact with an ore body whereas roots on the other side were in barren ground. Selection of vegetation from the wrong side of the plant could obviously produce unfortunate results.

A novel approach to biogeochemical prospecting and one which is connected with root systems, is the so-called *tree ratio method* suggested by Kleinhampl and Koteff (1960). This procedure is dependent on the different moisture requirements of the pinyon pine and juniper, both of which are phreatophytes.

Uranium in the Circle Cliffs area of Utah is found in sandstone-filled channel scours in Triassic and Jurassic rocks of the Colorado Plateau. By virtue of their porosity and permeability, these sandstone formations contain more available water than the adjacent mudstones and siltstones. The ratio of the abundance of the pinyon pine (high water requirement) to that of the juniper (low water requirement) was compared with the known depth of sandstone in 7 test areas. The results are shown in Figure 11-7. It can be seen that there is an approximate relationship between the two variables. The greater depth of sandstone indicates a greater probability of finding uranium in the area.

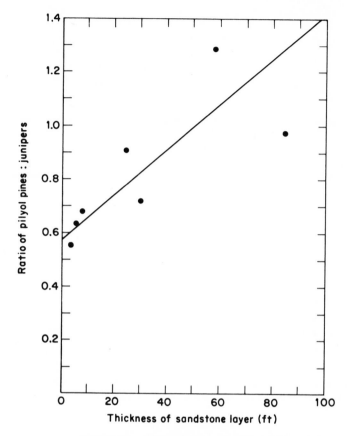

FIGURE 11-7 Abundance ratios of pinyon pines and junipers as a guide to depth of sandstone on the Colorado Plateau. Data from Kleinhampl and Koteff (1960).

DRAINAGE   Drainage is another factor that can affect prospecting by plants. Poorly drained soils when compared with those which are well drained can give misleading anomalies during an exploration project because many elements become more mobile under waterlogged conditions and this will be reflected in greater uptake by plants. Mitchell (1955) has studied this factor and carried out extensive trials on waterlogged soils. He found a greatly increased mobilization of cobalt, copper, iron manganese, nickel, and vanadium as measured by extraction of the elements into acetic acid and ammonium acetate. His data are presented in Table 11-9. Similar results for manganese in soils were obtained by Fujimoto and Sherman (1948).

Although increased elemental uptake due to drainage problems may not be as great as that caused by a genuine anomaly, any apparent anomalies associated with a change of the soil moisture content should be examined with care and reassessed critically.

TABLE 11-9   Extractability of Various Elements in Well-Drained and
Poorly-Drained Soils[a]

Soil drainage	Extractant	Extractability (%)					
		Co	Cu	Fe	Mn	Ni	V
Good	Acetic acid	0.52	0.21	20	—	0.36	0.09
Poor	Acetic acid	2.50	1.40	500	—	4.20	3.80
Good	Ammonium acetate	0.02	—	5	0.10	0.02	0.01
Poor	Ammonium acetate	0.37	—	5	1.60	0.44	0.03

[a]Depth: 9–15 in.

SOURCE: After Mitchell (1955).

ASPECT   The question of aspect is of paramount importance in plant ecology. It also has significance in biogeochemical prospecting because accumulation of elements from the soil is controlled by photosynthetic processes. Uptake will therefore be influenced to some extent by the intensity and duration of the light received by the plant. Pot trials such as those of Ferres (1951) have tended to confirm the influence of the photosynthetic process on mineral uptake.

The daily variations in rainfall received is also a factor that can influence mineral uptake by plants. Warren, Delavault, and Irish (1952b) have shown that dry soils can increase the iron content of plants and decrease the manganese concentration. The combined effects of soil temperature and moisture content are illustrated by the findings of the same workers that, in many cases, the metal contents of the same plant species differ markedly between sites of north and south aspect.

AVAILABILITY AND ANTAGONISTIC EFFECTS OF ELEMENTS IN SOILS
The question of the availability of elements in soils has already been discussed at some length and will not be mentioned further beyond reaffirming its importance as a factor influencing the success of biogeochemical methods of prospecting.

The question of mutual antagonism of pairs of elements is also a factor which should be considered. In work in New Zealand, the author has frequently noticed that if the concentrations of pairs of elements in vegetation are plotted against each other, a linear plot of negative slope is obtained. Figure 11-4 shows a plot of lead values of a trunk section of the New Zealand species *Beilschmiedia tawa* plotted as a function of the copper content. The inverse relationship is quite apparent.

Similar inverse relationships have been found for nickel-calcium, nickel-magnesium, cobalt-calcium, and calcium-magnesium accumulated in serpentine plants of New Zealand (Lyon, 1969). The reason for this behavior is not clear; it may be linked to a breakdown of the plant's selection mechanism. The capacity of a plant to absorb mineral salts obviously has finite limits. If one constituent of the soil is accumulated to an excessive degree, this will have to be compensated by reduced uptake of another.

11-8   AN OVERALL ASSESSMENT OF FACTORS AFFECTING ELEMENTAL ACCUMULATION BY PLANTS

In the previous section of this chapter, some 12 variables affecting elemental uptake by plants have been considered and have been discussed under nine separate subheadings. The list is not complete and there are a number of other factors concerned with the soil which can also affect the technique. One of the most surprising features of biogeochemical prospecting is that it can apparently be used successfully in spite of all the numerous factors which can affect the reliability of the method. The inference is clear. Many of the variables must have a minor effect, though in some areas they can be more important than elsewhere. Other more important variables may be reduced or even eliminated completely by use of certain procedures.

A successful technique for a reduction of interferences in biogeochemical prospecting has been the use of a ratio of a pair of elements as an indication of mineralization (Warren and Delavault, 1949; Worthington, 1955; Tooms and Jay, 1964). This procedure involves measuring the concentration of the element being sought and also that of another element used as a control. For example, the ratio copper/zinc in vegetation may well give an indication of copper mineralization provided that zinc mineralization does not occur simultaneously. The assumption is made that all factors affecting the uptake of copper from the soil also affect zinc to the same extent and hence are cancelled out when the ratio is used. The copper/zinc ratio is not a particularly good one to use for reasons explained in Chapter 14 but ratios such as nickel/cobalt are quite suitable and are used extensively at present in the Soviet Union in the search for nickel (Malyuga, 1964).

Table 11-10 lists some of the factors affecting the accumulation of elements by

TABLE 11-10   Factors Affecting Elemental Uptake by Plants and Methods of Reducing Their Effect

Factor	Relative importance	Method of reducing effect
Type of plant	Very great	Selection of plants by orientation survey
Organ sampled	Great	Selection of organ by orientation survey
Age of organ	Significant	Selection by orientation survey
Root depth	Significant	Use of ratios of two elemental concentrations in same sample
pH	Fairly significant	Selection of elemental ratios of pair of elements of same availability over pH range encountered
Health of plant	Fairly significant	Select healthy specimens only
Drainage	Fairly significant	Avoid poorly drained areas where possible; also use elemental ratios
Availability of element	Fairly significant	Use elemental ratios with both elements of same availability
Antagonism of other elements	Minor	None
Rainfall	Minor	Carry out work over short period
Variable shading	Minor	Avoid shady sites if possible or use elemental ratios
Temperature of soil	Minor	Carry out work over short period

plants in order of decreasing general importance. Suggested procedures for reducing the effects of these variables are also included. It is not claimed that the measures recommended in Table 11-10 will completely eliminate the adverse effects of all the variables concerned, but judicious use of them should help to ensure that any variations which do occur are less than the real variations caused by the presence of anomalies in the exploration area.

SELECTED REFERENCES
General: Carlisle and Cleveland (1958)
Plant Nutrition: Sutcliffe (1962)

CHAPTER TWELVE
BIOGEOCHEMICAL
ORIENTATION
AND
EXPLORATION
SURVEYS

An orientation survey is an essential prerequisite for biogeochemical prospecting when new plant species are to be used for the first time. It is often supposed that the necessity for such a survey discourages use of the biogeochemical method if soil sampling can be used instead. This view presupposes that orientation surveys are not needed in soil sampling; but this is far from the truth since, in most cases, the field worker has to study elemental distributions in the soil profile in order to determine which soil horizons are the most likely to represent bedrock values or mineralization.

The following discussion concerns the recommended procedures for carrying out biogeochemical orientation and exploration surveys and includes the necessary basic information.

12-1 SELECTION OF A SITE FOR AN ORIENTATION SURVEY

In selecting a site for an orientation survey, the field worker should be guided by a number of important considerations. The first of these is that the area should contain an anomaly or anomalies. It is not always necessary that the anomaly itself should be an economic ore body, but it should contain the element or elements that are sought, at levels at least two or three times those of normal background levels. Concentrations of elements in anomalies of the test area should ideally approximate those of the area which is to be surveyed later. Since, however, information on the exploration region is seldom known beforehand, it will only be apparent later whether the anomalies of the orientation survey were well chosen or not. The apparent success of the orientation survey will depend largely on the contrast between background and mineralization in the test area and it will therefore be desirable that anomalies in the exploration area have equal or even higher elemental concentrations than those of the test region. Under these circumstances, the orientation-survey assessment of the potentiality of the biogeochemical method will be conservative. The assessment will be overoptimistic if mineralization in the exploration area is weaker than in the test area.

A second consideration for the choice of site is that it should be climatically, ecologically, geologically, and geomorphologically representative of the whole area which is to be surveyed ultimately.

The area of the survey region should be such that it would allow for about 100 sampling points at spacings which would be employed for the main exploration work itself. Assuming a grid pattern of sites situated 100 ft apart, this would allow for an area of less than a tenth of a square mile.

Absence of known anomalies in completely virgin territory presents another problem. To some extent this can be overcome by selecting as a test site a contact between two different geological formations in which the background concentrations of the elements sought might be appreciably different. If no contacts of formations of different elemental concentrations are to be found, it is still possible to use the large amount of data which is slowly being accumulated (see Section 12-2) concerning normal background levels of elemental abundances in various species of plants. In New Zealand, biogeochemical prospecting for copper and nickel has been done by using data from an orientation survey carried out with the same species on formations geologically similar but situated some 300 miles away from the exploration area.

12-2   SELECTION OF PLANTS FOR TESTING

Although virtually *all* the vegetation in an intrazonal biogeochemical province will give some indications of mineralization, great care must be taken in selecting the most favorable species from the plant assemblage. The field worker should first visit the area and determine the abundance and distribution of large plant species (see Chapter 3).

The abundance of plants is of great importance since it is essential that those selected for the survey should be obtainable at most sampling sites throughout the whole area; otherwise, there will be gaps in the sampling network. Infrared photography (Chapter 6) is an excellent way of determining the abundance and distribution of plant species. Several species should be selected for further tests after the application of the abundance criterion, in order to ensure that there will be at least one of the species at each sampling point.

There are other considerations besides abundance in the selection of plants for further testing. Care should be taken that the species are not too large. A very large tree will often not have leaves or twigs within easy reach of the ground or those that are available may be light-starved, spindly growths atypical of the main branches further up the trunk. The opposite extreme is also to be avoided. Trees or shrubs that are too small are unlikely to sample an appreciable amount of soil and will have little penetrating power through the overburden. A further complication of the small volume of soil sampled is that the elemental concentrations in the plant will tend to have a wider scatter than in the case of larger plants which can sample more soil.

Application of the above selection criteria will leave the field worker with a small number of species for further testing. If this number is still appreciable, it may be reduced further by carrying out a "minisurvey" involving sampling all the species and their associated soils at about ten sites in a traverse of an anomaly. By plotting plant and soil values on a graph, it will be possible to eliminate visually the less promising species.

12-3   SELECTION OF A SAMPLING PATTERN

The selection of the sampling pattern will be dictated mainly by the topography and the size and shape of the anomaly or anomalies in the area. The smaller the size of the dispersion halo or anomaly, the closer together the traverse lines will have to be. A square grid or rectangular grid pattern is desirable since these are easy to survey. The traverse lines should be at right angles to the strike of the ore structure and should be so arranged that at least two points fall within important anomalies. It is preferable that the grid pattern be so arranged that about one third of the points will lie on background, one third on moderately mineralized ground, and the remainder on anomalous terrain. Where the ground is extremely steep and dissected, it may be advisable to rely instead on a pattern following spurs and ridges. In this way there will be more likelihood of all samples being obtained from sites of similar aspect and drainage. The criteria for determining the pattern of sampling sites for vegetation are exactly the same as for soil surveys. The reader is referred to Hawkes and Webb (1962) for further discussion of this subject.

**12-4   METHODS OF SAMPLING AND STORAGE OF
PLANT AND SOIL MATERIAL**

During the orientation survey, the field worker is advised to collect samples of leaves as well as young twigs from the previous year's growth. This material is young enough to have gone through the main processes of mineral accumulation and yet is not so old that senescence, with an attendant possibility of loss of minerals, has started to occur to any extent. Other plant organs such as flowers, fruits, and bark may also be collected, but experience has shown that the best results will probably be obtained with leaves or twigs (Lyon and Brooks, 1969). The question of sampling in biogeochemistry has been investigated extensively by Warren and his co-workers (Warren and Delavault, 1948; Warren, Delavault, and Irish, 1952a; Warren, Delavault, and Fortescue, 1955). The reader is referred to these papers for further information.

The amount of vegetation collected should be sufficient to provide about 10–30 g of dry material; this would be provided by about 100 g of fresh vegetation. The samples should be removed from the trees with pruning shears or hedge trimmers (for tall species), placed in prenumbered plastic bags, and sealed with elastic bands. The samples should be collected at various points around the circumference of the tree or shrub and preferably at a point as high as can be reached conveniently from the ground.

Soil samples should be taken from various points around the base of each plant sampled and should be taken from the appropriate horizon below the loose surface humus. A total of about 200–250 g should be adequate. The samples may be placed directly into kraft paper bags or finely woven cotton sacks.

Plants and soils should be collected over the whole of the predetermined grid and, as far as possible, the operation should be carried out over a limited number of days.

At each sampling point, observations on the topography and geology of the site, type of vegetation, existence of outcrops, etc. should be entered in the field book. It is usually better to take prenumbered bags at random from a container and enter this number in the field book for the appropriate site rather than attempt to number bags in the field. Under adverse weather conditions, freshly marked numbers can soon become illegible. This also applies to the field book itself and in wet areas it is advisable to construct a book from the tracing plastic used in cartography.

The nature of subsequent treatment of the samples will depend on whether they are to be analyzed in the laboratory or in the field. Generally, any operation in the field is considerably more difficult and less reliable than one carried out in the laboratory. Unless weight is a problem, it is better to take the samples back to the laboratory for further processing. Vegetation samples are initially collected in sealed plastic bags to keep them fresh before washing, which is the next stage of the process. Leaves stored in plastic bags will keep for about a week in a fairly fresh condition provided that temperatures do not rise much above 60°F. Twigs will of course keep longer. Unless refrigeration facilities are available at the base camp, vegetation samples will have to be further processed in the field if their dispatch to the laboratory is delayed beyond a week.

On return to the laboratory, soil samples are laid out on numbered pieces of paper and are air dried for about two days. Alternatively, the paper or cotton bags of samples may be dried in an oven at 110°C. After drying, the soils are lightly disintegrated with a mortar and pestle and are then sieved through a 15-mesh nylon sieve before storage for further treatment. In arid areas, soils may be sieved directly *in situ*.

When vegetation is removed from the plastic bags, it should be washed vigorously under running water with a final rinse in distilled water. The samples should then be dried at 110°C in a drying oven. If the vegetation consists of leaves, crush the material firmly in the hands after drying. This will remove most of the leaves from the supporting stems and small twigs. The material may then be placed in bottles or plastic bags until required again. Twigs should be cut into small sections with pruning shears *before* the drying process. Once the material is dry, it is very much harder to cut.

The foregoing assumes that materials are to be taken to the laboratory for analysis. If field analyses are to be carried out or if samples are to be reduced in weight for onward transmission, a different procedure should be followed. The soil bags should be suspended in a warm dry place at the base camp until dry. The samples are then gently disintegrated by hand or in a mortar and pestle and are sieved as before. The bulk of the samples is reduced by coning and quartering and a small representative sample is prepared for onward dispatch.

In the field, vegetation samples should be washed in a stream or water tank and then transferred to paper bags for drying in the sun. This procedure is not always possible since, in many areas, sun or water or both are lacking. If water is not available, the washing procedure has to be eliminated. This may cause contamination problems (see Section 12-7) which can be overcome in the case of twigs by peeling the specimens, a step that is laborious in the extreme and should not be contemplated except in cases of dire necessity.

As a further step, the vegetation may be partially calcined in the field by heating in aluminum pots held over a primus stove. There will be no need at this stage to ash the material completely since this can be carried out in a proper muffle furnace at a later stage. The calcined material is then pulverized in the original container by means of a glass rod and stored in plastic bags or bottles until required.

12-5   FINAL PREPARATION OF MATERIAL FOR ANALYSIS

PREPARATION OF VEGETATION SAMPLES   Assuming that the soil and vegetation samples are to be analyzed by either emission spectrography or by methods such as atomic absorption spectrophotometry (see Chapter 15) which require a solution, the procedure for further treatment of the samples will usually involve "dry" ashing in a muffle furnace. This step is absolutely essential for emission analysis and is more preferable than "wet" ashing when "solution" chemical methods are to be used.

Due to the limitations imposed by furnace space, the main problem in calcining plant material is to strike the correct balance between the amount of material needed to give a representative sample and the amount which can be economically handled to give a fast output from the laboratory. One way to improve the economics of the method is to pulverize a quantity of the dried material in a micro-hammer mill, thus producing a finely powdered product from which a much smaller representative sample can be obtained. Even a small automatic coffee grinder can be quite effective for this purpose. Assuming that a microhammer is used, only about 10 g of dried material will be needed and this is placed in a 25-ml squat Pyrex beaker and ashed in a muffle furnace at 450°C for several hours. This will produce about 0.2 g of ash from woody material and about 0.5 g from leaves. An alternative procedure is to heat leaves or twigs in 150-ml beakers over a hot plate. Charring will ensue and the residue will fit a 10-ml squat beaker for subsequent ignition in the furnace. Both methods will produce an amount of ash sufficient for the determination of a number of elements. The average muffle furnace will take about 50 beakers of the 25-ml size and this number can be more than doubled if 10-ml beakers can be used. Assuming that the muffle is left running continuously, 150–200 samples can be ashed every 24 hours.

Ashing should be carried out in the presence of a small amount of air. If too much air is admitted, the material will catch fire; if too little is permitted access, there is a danger of volatilization of some constituents. Suddenly admitting air after a period of closure of the oven door usually results in an explosion if organic material is still present.

There has been much controversy concerning the amounts of elements lost during dry ashing. Most workers agree, however, that at 450°C little loss will occur (Mitchell, 1964). At 500°C ashing is effected more quickly, but there is the risk of losing volatile constituents such as cadmium and lead at this temperature. Some workers prefer wet ashing of plant material with a mixture of perchloric and nitric acids (see Chapter 15), but this operation is expensive and dangerous and in any case is not suitable if emission spectrography is to be used as an analytical tool.

PREPARATION OF SOILS   If soils are to be analyzed by solution methods, little further treatment is required of the 15-mesh material beyond further grinding with an agate mortar and pestle until the soil passes a 60-mesh sieve. Some workers prefer to ash the 15-mesh soil at 450°C before analysis by wet methods and then grind the calcined material to 60-mesh size. If emission spectrography is to be used, this step is absolutely essential anyway. When igniting soils, the recommended procedure is to place about 10 g of the material in the 25-ml squat beakers and proceed as for plant material.

12-6   THE WEIGHT BASIS OF EXPRESSING ANALYTICAL DATA
After the samples have been analyzed, the data may be expressed either on the basis of the dry weight (110° C) or the ashed weight. For vegetation, this makes a difference of about ×50 for woody material and ×20 for leaves. Malyuga (1964) and Warren et al. (1955) recommend that the data be expressed on an ash-weight basis and this appears to be the most generally accepted method of expression of results. Provided one system or the other is followed throughout, little confusion will arise.

The treatment of soils presents an entirely different problem. Most normal soils free of loose surface humus contain about 10% humic material or less and therefore comparatively little error will arise whether the results are expressed on a dry-weight or ash-weight basis. If, however, soils are naturally dark in color, there is a risk of inadvertently including extra humic material and great errors can be introduced by expressing the data on an ash-weight basis. This is illustrated in Figure 12-1 which shows a traverse of a nonmineralized area in New Zealand. The copper and nickel content of the plants does not vary appreciably over the region and neither does the elemental concentration of soils expressed on a dry-weight basis. When, however, the soil values are expressed on an ash-weight basis, a false anomaly immediately becomes apparent. The author therefore strongly recommends that all soil values be expressed on a dry-weight basis whether the material was ignited or not.

12-7   CONTAMINATION PROBLEMS
The risk of contamination of vegetation from natural causes during a biogeochemical survey varies with a number of factors. The most common problem of all is wind-blown dust from mine workings, smelter fumes, or exposed soil. In New Zealand, with its dense forest cover and high rainfall, contamination effects of this nature are rare and have never been a serious problem in field work. In arid regions, however, contamination can be a serious difficulty.

Cannon (1952) has shown some of the contamination effects which can arise in vegetation. Some of her data are shown in Table 12-1 and are self-explanatory. The same author (1960a) has stated that ore trucks can add 1–2 ppm uranium to

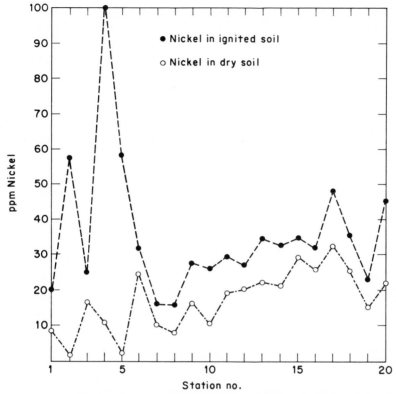

FIGURE 12-1   Nickel values in humus-rich soils sampled during a traverse of the Longwoods Range, South Island, New Zealand. The figure illustrates the importance of expressing soil analyses on a dry-weight rather than ash-weight basis. Due to the high organic content, the value for nickel at station 4 gives a false anomaly when expressed on an ignited weight basis.

vegetation and that a mill can increase the content of this element in trees by a factor of 1000.

Another form of contamination is from nonmineralized dust which has the effect of diluting the plant ash and hence lowering concentrations of some elements. This factor is, however, except in the case of bryophytes, fairly negligible.

One of the most serious contamination effects is that which originates from the tetraethyl lead of automobile exhaust fumes. Unlike other forms of contamination, the lead is so strongly absorbed by vegetation that 80% of it still remains after washing (Warren and Delavault, 1960). Pollution of the biosphere by tetraethyl lead is rapidly becoming a world-wide problem of serious dimensions, particularly in the United States. The author himself has even detected lead contamination in ocean waters off the Los Angeles Basin at a distance of over 30 miles from land (Brooks et al., 1967).

Warren and Delavault (1960) were the first to report abnormal concentrations of

TABLE 12-1   Contamination Effects on Plants Growing in Unmineralized Ground Near a Uranium Mill and Mine

			ppm U in ash	
Species	Location	Time	Washed	Unwashed
Cowania mexicana	Near mine	During operations	—	51.0
Cowania mexicana	Near mine	After mine closed	6.4	8.4
Juniperus monosperma	Near mine	During operations	—	7.8
Juniperus monosperma	Near mine	After mine closed	5.3	7.0
Juniperus monosperma	Adjacent to mill	—	—	1100.0
Juniperus monosperma	800–1500 ft from mill	—	—	150.0
Juniperus monosperma	2000–4000 ft from mill	—	—	40.0

SOURCE: Cannon (1952).

lead in vegetation situated near highways.   Values in plants were as high as 3100 ppm in one case and averaged 1200 ppm for 4 species studied.   Mean background levels for the same species were only 40 ppm.   These data are shown in Table 12-2.   Further observations on lead pollution in vegetation have been made by Warren (1966) and Cannon and Bowles (1962).   The latter workers showed that, on average, lead contamination from automobiles was apparent in grasses at distances of up to 500 ft from major highways.   These data are shown in Figure 12-2.

The above discussion has clear lessons for biogeochemical prospecting.   Surveys involving lead should, wherever possible, be carried out in areas at least 500 feet away from a major highway.   Any lead "anomalies" found near highways should be treated with some suspicion until investigated more thoroughly.   Since the zinc content of mineralized soils is very often related to the lead concentration, one way of testing the validity of a lead anomaly indicated by plants will be to analyze zinc as well.

There are other forms of contamination of vegetation apart from that of automobile exhaust.   Top-dressing with fertilizers or the spraying of vegetables with insecticides or fungicides containing inorganic constituents can be a source of error in biogeochemical investigations and should be taken into account where necessary.

TABLE 12-2   Lead Concentrations in the Ash of Twigs of Trees Growing Close to Highways Compared with Normal Background Values

		Lead content (ppm)			
		Near highways		Normal background	
Species	No. of samples	Range	Mean	Range	Mean
Pseudotsuga menziesii	10	510–2200	1200	14–60	38
Thuya plicata	7	630–3100	1600	23–74	51
Abies grandis	4	350–1100	800	23–56	39
Betula glandulosa	27	270–2500	1300	21–53	32

SOURCE: Warren and Delavault (1960).

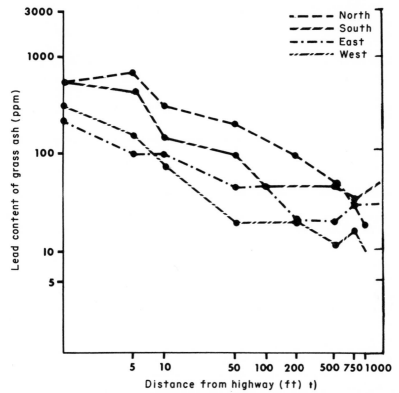

FIGURE 12-2 Variations in the lead content of the ash of grasses marginal to two major highways in Denver, Colorado. Values shown for distances of up to 1000 feet from either roadway. From Cannon and Bowles (1962). Copyright 1962 by the American Association for the Advancement of Science.

12-8   PATHFINDER ELEMENTS

The concept of *pathfinders* is well established in mineral exploration, particularly in geochemical work. Pathfinders may be defined as elements which, because of some property or properties, provide anomalies or halos more readily usable than the element sought and which are geochemically associated with that element. Warren and Delavault (1959) have described several successful uses of pathfinders in mineral exploration and many other cases have been reported in the literature.

Warren, Delavault, and Barakso (1964) have shown that pathfinders may also be used in biogeochemical prospecting. The most useful species tested was the Douglas fir (*Pseudotsuga menziesii*). The normal arsenic content of growing tips of this species is less than 1 ppm. In areas of sulfide mineralization where arseno-pyrite is an important constituent, amounts of up to 10,000 ppm (ash weight) were recorded in the vegetation. The data are shown in Table 12-3. It will be seen that, with one exception, the biogeochemical anomaly is far greater than the soil

TABLE 12-3   The Arsenic Content of Soils and Growing Tips of Douglas
Fir Trees from Mineralized Regions of British Columbia

District	Arsenic in vegetation (ppm)	Arsenic in soils (ppm)	Plant/soil ratio
Bridge River	10,000	4,633	2.2
Similkameem	8,800	2,980	3.0
Bridge River	2,500	1,072	2.3
Bridge River	50	218	0.2
Bridge River	1,550	39	39.7
Salmo	560	38	14.7
Kimberley	17	14	1.2
Bridge River	33	9	3.7

SOURCE: Warren, Delavault, and Barakso (1964).

anomaly. Arsenic in soils and rocks already has a reputation of being a good pathfinder for gold and sulfide minerals so that the possibility of using it for this purpose in biogeochemical prospecting is of great interest.

Warren, Delavault, and Irish (1952b) have investigated the use of iron and manganese in vegetation as indicators for other elements. Tkalich (1953) has used the iron content of plants for the same purpose. Talipov et al. (1968) have explored the possibility of using arsenic in vegetation as a pathfinder for gold and obtained positive results.

The use of pathfinders in vegetation is a possibility to be explored during an orientation survey. Such data are usually obtained from the statistical analysis of interelemental relationships in vegetation itself, a procedure which should always be undertaken at the end of the orientation survey.

12-9   EVALUATION OF AN ORIENTATION SURVEY

Evaluation of an orientation survey will only be considered very briefly in this section because much of the statistical treatment will be discussed in the next chapter.

After collection of samples from an orientation survey and their preparation for analysis, the next steps will be chemical analysis (Chapter 15) and statistical analysis (Chapter 13). On completion of these two steps, the field worker will have acquired several items of important information which will guide him in subsequent exploration work in unknown areas. The first of these is information on the degree of correlation of the elemental content of the various plant species with the corresponding values in the soils or bedrock. To evaluate this relationship, the *correlation coefficient* (*r*) is obtained (see Chapter 13). This function varies in value between +1 and −1. A value of +1 for two sets of data implies that there is a completely linear relationship between them and that they are directly proportional to each other. A value of −1 still implies the perfect linear relationship but shows that the two variables are inversely related. As values of *r* approach 0 from either side, the trend toward nonsignificance increases.

A second item of great importance to the field worker is the determination of what concentration of an element in plant ash shall be taken as representing an anomaly in the soil or rock beneath.

The statistical treatment should provide information not only on the best species to select for exploration surveys, but also as to which organ should be selected and whether the data should be expressed on a dry-weight or ash-weight basis.

An indication of the reproducibility of element accumulation in a plant is provided by calculation of the coefficient of variation (Chapter 13) for the elemental content of different specimens of the same species selected from an area with a uniform content of that element.

Emphasis in this chapter has been on plant–soil relationships. A more meaningful evaluation of the potential of vegetation for use in prospecting would be afforded by comparison of plant data with *bedrock* values of the element which is sought. This procedure is seldom carried out because of the difficulty of obtaining a sufficient number of bedrock samples to make a project statistically meaningful. If sufficient samples can, however, be obtained, a very useful exercise is to compare plant–soil data not only with themselves but separately with bedrock values.

Another approach to the orientation survey is a comparison of plant values for a particular element with data obtained by geophysical techniques such as *induced polarization* or *geomagnetic* measurements. Again, little work of this nature has been carried out. A notable exception is the work of Marmo (1953) who compared hydrological and geomagnetic data with plant values for copper and zinc in Finland.

12-10 BIOGEOCHEMICAL EXPLORATION SURVEYS

With the completion of the orientation survey, the field worker will now wish to apply the methods to virgin territory. In some areas only soil surveys would be feasible (i.e., in deserts), in others the most effective procedure might be biogeochemical work (see below), but in the majority of cases a joint plant–soil survey would be advantageous.

To decide the relative merits of soil or vegetation sampling, it is advisable to expose the soil profile by digging a pit and to sample various horizons for analysis of their constituents. If the element that is sought is concentrated in one of the upper horizons, it is possible that there would be no particular advantage in using plants as a guide to mineralization, since soil sampling would be very easy. If the element is enriched at depth, however, and particularly if there is an impenetrable geochemical barrier such as a siliceous hardpan or carbonate layer between the surface and zone of enrichment, plants will probably be useful for prospecting. This is demonstrated in Figure 12-3, which shows the distribution of nickel in a soil profile containing a carbonate-rich surface layer. Detection of the nickel anomaly by soil surveys would be dependent on sampling either the loose humus layer or the C horizon at depth. Most trees and large shrubs should be capable of penetrating to the desired depth in this soil profile and the value of plants for this purpose is obvious.

Other situations in which the biogeochemical method could be used along with soil surveys, or even in place of them, are in cases where the terrain is very steep and eroded so that transported anomalies in soils are to be found. Newly colonizing deep-rooted plants could presumably sample through to bedrock and give

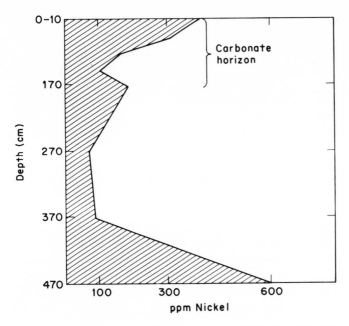

FIGURE 12-3   Distribution of nickel in a soil profile with a carbonate upper horizon.   From Malyuga (1964).   By courtesy of the Plenum Publishing Corp.

evidence of mineralization, whereas the corresponding soil anomaly would appear some distance downslope. It must be admitted, however, that plants growing in the transported anomaly itself will probably give evidence of mineralization here also.

In terrain with stable soils and gentle relief and where elements are enriched in the more accessible soil horizons, there may be little or no advantage in using the biogeochemical method. Nevertheless, for the sake of completeness and because no one method of exploration is complete in itself, it is probably always advisable to include both soil and vegetation surveys whenever economic considerations make this possible.

The working procedure in an exploration survey is little different from that of an orientation project except that probably fewer plants will be sampled. It may be advisable to sample several species throughout the operation in order to ensure that at least one species will be present at every sampling site.

12-11   INTERPRETATION OF AN EXPLORATION SURVEY

When the vegetation samples have been analyzed, the first stage is to plot the data on a map of the area. As it is envisaged that more than one species will have been sampled, the elemental concentrations should first be reduced to a common standard by dividing concentrations of elements in plant material by the enrichment coefficient (see Chapter 11) of the plant for each element investigated.

Enrichment coefficients will have been determined previously from an orientation survey or will have been derived from the literature.

There are many pitfalls in plotting contour maps of elemental isoconcentrations in plants or soils. Not the least of these is the fact that anomalies of different shape can be produced merely by selecting a particular threshold level for elemental concentrations in plants or soils. To avoid this difficulty, it is essential to use as a guide (though not invariably) the threshold level indicated by the statistical analysis of data from an orientation survey. Once this value has been obtained, it is still possible to make serious errors in drawing isoconcentration contours. Figure 12-4 shows a correct and incorrect method of drawing these contours. The illustration is part of a rectangular grid pattern which is easiest to handle.

The map is first divided into units of four grid rectangles and then faint pencil lines are drawn from the center of each unit to its eight nearest neighbors. The concentration gradient along each line is calculated and the desired threshold level of the isoconcentration contour is inscribed in the appropriate position.   These positions are then joined together with a curve. In practice, plotting contour lines in this manner is extremely laborious but the work can be much simplified by the use of computers equipped with automatic plotting facilities. Such facilities are now available in some exploration companies and in many analytical service laboratories.

Figure 12-5 shows a contour map depicting the outline of a nickel anomaly in New Zealand as determined from plant and soil analyses.

Contour maps can also be plotted using elemental ratios of plants and soils such as the following pairs: copper–zinc, nickel–cobalt, thorium–uranium, uranium–radium. Use of such ratios carries the advantage that the use of enrichment coefficients is eliminated and so are many of the variables which can affect elemental uptake by plants. Elemental ratios present in ore bodies are usually preserved in

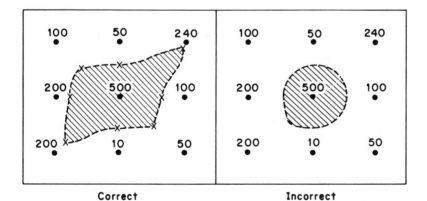

FIGURE 12-4   Representation of correct and incorrect representation of contour lines for isoconcentrations (250 ppm) of an element in soils or vegetation situated over an exploration target. Shaded areas indicate region assumed to contain more than 250 ppm of the element concerned.

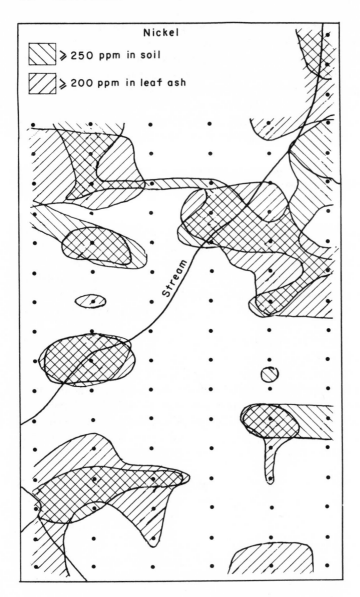

FIGURE 12-5 Degree of coincidence of nickel anomalies in the Riwaka Basic Complex, Nelson Province, New Zealand, as determined by analysis of soils and plant ash of *Nothofagus fusca.* From Timperley et al. (1970a). By courtesy of the Economic Geology Publishing Co.

the dispersion halo and are reflected in some plants in a different, although reproducible, manner.

Examination of the contour map obtained from a biogeochemical exploration survey will show anomalies common to both plants and soils and some exclusive to either plants or soils separately. Where anomalies are confirmed by the combined data, extra confidence may be placed in the site as being favorable. Other areas, however, should not be dismissed as being nonsignificant when agreement between plants and soils is not evident. A reason should always be sought for such unsupported anomalies.

Anomalies common to plants only, particularly in steep terrain, may indicate absorption of soluble salts washed down from a true anomaly situated uphill or may instead indicate bedrock mineralization from which the original overburden has been transported. Exclusive plant anomalies shown over boggy soils, however, may be due to inadequate drainage (Chapter 11).

Anomalies common to soils only may be due either to a true anomaly or to the soils themselves being part of a transported anomaly.

12-12   BACKGROUND DATA FOR BIOGEOCHEMICAL PROSPECTING

As a usable pool of knowledge of normal background concentrations for elements in certain plant species is accumulated, the need for extensive orientation surveys becomes progressively reduced. When the pool of knowledge includes common species of continental distribution, it can be of great use for biogeochemical prospecting. A very large proportion of the existing knowledge of background data is derived from North America, mainly because of the pioneering work of Warren (1962). Available data for the Soviet Union are much more sparse, because, although the amount of information acquired by that country on background levels of elements in plants must be many times greater than in the rest of the world combined, much of the work has remained unpublished or else has appeared in publications not translated or abstracted in the West.

Table 12-4 is a compilation of data from various sources and shows the normal background levels of a group of elements in the ash of certain plants. The table

TABLE 12-4   Normal Background Elemental Concentrations in Plant Ash

Species	Common name	No.	Cu	Mo	Ni	Pb	Zn
					Elements (ppm)		
ALASKA (Shacklette, 1960, 1962b)							
Gaultheria shallon (S + L)	Salal	27	—	—	—	90	1100
Menziesia ferruginosa (S)	Menziesia	76	—	—	—	190	2600
Menziesia ferruginosa (S + L)	Menziesia	61	—	—	—	80	1600
Picea glauca (S + L)	White spruce	—	40	—	—	25	1500
Picea mariana (S + L)	Black spruce	—	80	—	—	25	2000
Picea sitchensis (S + L)	Sitka spruce	59	150	—	—	120	1300
Tsuga heterophylla (S + L)	W. hemlock	39	—	—	—	150	1100
Vaccinium ovalifolium (S + L)	Early blueberry	45	—	—	—	70	1500
Vaccinium parvifolium (S + L)	Red huckleberry	10	—	—	—	40	1600
AUSTRALIA (Nicolls et al., 1965; Cole et al., 1968, and author's unpub. data)							
Acacia aneura (L)	—	21	70	—	—	—	—
(S)	—	40	39	—	—	—	—
Acacia linophylla (L)	—	25	43	—	—	—	—
(S)	—	25	36	—	—	—	—

TABLE 12-4 (Continued)

Species	Common name	No.	Cu	Mo	Ni	Pb	Zn
					Elements (ppm)		
Acacia salicina (L)	—	6	40	—	—	—	—
(S)	—	6	30	—	—	—	—
Cassia helmsii (L)	—	8	81	—	90	—	237
(S)	—	6	85	—	37	—	165
Cassia nemophila (L)	—	10	52	—	44	—	257
(S)	—	6	53	—	35	—	102
Cassia sturtii (L)	—	21	61	—	70	—	241
(S)	—	10	69	—	28	—	139
Eriachne mucronata (L)	—	3	60	—	—	<10	170
Gomphrena canescens (L)	—	2	19	—	—	10	205
(S)	—	2	48	—	—	40	230
Polycarpaea corymbosa (L)	—	1	85	—	—	30	300
(S)	—	1	87	—	—	5	950
Polycarpaea spirostylis (L)	—	3	42	—	—	<20	880
(S)	—	3	200	—	—	460	600
Tephrosia sp. (L)	—	4	55	—	—	<10	220
Trianthema rhynchocalyptra (L)	—	1	30	—	—	20	180
Triodia pungens (L)	—	6	47	—	—	<10	178
	CANADA (Warren, 1962)						
Abies amabilis (S)	Balsam	5	130	—	—	—	1300
Abies grandis (S)	Balsam	10	160	—	—	—	1200
Abies lasiocarpa (S)	Balsam	7	220	—	—	—	1400
Acer circinatum (S)	Maple	3	160	—	—	—	440
Acer glabrum (S)	Douglas maple	16	130	—	—	16	770
Acer macrophyllum (S)	Maple	3	160	—	—	13	900
Alnus rubra (S)	Red alder	3	390	—	—	40	1300
Alnus sinuata (S)	Sitka alder	6	230	—	—	9	1200
Alnus tenuifolia (S)	Mountain alder	2	300	—	—	22	1300
Amelancher alnifolia (S)	Saskatoon berry	11	180	—	—	20	1100
Apocynum androsaemifolium (S)	Dogbane	1	160	—	—	—	310
Arctostaphylos uva-ursi (S)	Kinnick-kinnick	1	140	—	—	30	2000
Artemisia tridentata (S)	Sagebrush	4	190	—	—	—	280
Artemisia trifida (S)	Sagebrush	2	330	—	—	—	700
Betula fontinalis (S)	Water birch	6	220	—	—	40	4200
Betula papyrifera (S)	White birch	11	180	—	—	25	4600
Ceanothus velutinus (S)	Snowbrush	1	120	—	—	—	670
Chimaphila umbellata (S)	Pipsesse	2	90	—	—	12	1300
Chiogenes hispidula (S)	Snowberry	6	100	—	—	—	640
Chrysothamnus nauseosus (S)	Rabbitbush	1	280	—	—	22	630
Cornus stolonifera (S)	Dogwood	11	180	—	—	—	1000
Corylus californica (S)	Hazel	1	180	—	—	28	1400
Crataegus douglasii (S)	Black hawthorn	4	130	—	—	40	1000
Epilobium angustifolium (S)	Fireweed	3	100	—	—	—	1500
Equisetum sp. (S)	Horsetail	4	50	—	—	—	110
Gaultheria shallon (S)	Salal	7	120	—	—	—	900
Juniperus communis (S)	Dwarf juniper	10	140	—	—	22	470
Juniperus scopulorum (S)	Rocky Mt. juniper	4	150	—	—	—	520
Larix occidentalis (S)	Western larch	7	250	—	—	16	1700
Lonicera sp. (S)	Lonicera	2	180	—	—	—	1200
Mahonia aquifolium (S)	Tall oregon	2	220	—	—	10	750
Mahonia nervosa (S)	Grape oregon	7	250	—	—	—	950
Mahonia repens (S)	Creeping oregon	1	300	—	—	—	990
Malus communis (S)	Apple	3	140	—	—	21	230
Menziesia ferruginea (S)	Menziesia	5	450	—	—	—	4800

TABLE 12-4 (Continued)

Species	Common name	No.	Elements (ppm)				
			Cu	Mo	Ni	Pb	Zn
Oplopanax horridum (S)	Devil's club	9	130	—	—	—	510
Pachystymum myrsinites (S)	False box	2	180	—	—	16	1500
Philadelphus lewisii (S)	Mock orange	8	190	—	—	9	580
Picea engelmanni (S)	Engelman spruce	13	230	—	—	20	2300
Picea glauca (S)	White spruce	10	280	—	—	66	2000
Picea sitchensis (S)	Sitka spruce	1	400	—	—	160	1600
Pinus albicaulis (S)	Whitebark pine	5	230	—	—	—	1600
Pinus contorta (S)	Lodgepole pine	9	240	—	—	66	1500
Pinus monticola (S)	White pine	6	290	—	—	82	2000
Pinus ponderosa (S)	Ponderosa pine	14	150	—	—	60	1800
Populus tremuloides (S)	Aspen	10	210	—	—	—	2180
Populus trichocarpa (S)	Cottonwood	4	100	—	—	—	3100
Potentilla fruticosa (S)	Cinquefoil	1	280	—	—	33	900
Prunus cerasus (S)	Cherry	1	230	—	—	—	1000
Prunus demissa (S)	Choke cherry	11	140	—	—	25	730
Prunus emarginata (S)	Bitter cherry	2	120	—	—	66	750
Pseudotsuga menziesii (S)	Douglas fir	48	240	—	—	40	2200
Pteridium aquilinum (S)	Eagle fern	1	50	—	—	60	500
Purshia tridentata (S)	Antelope bush	1	160	—	—	70	600
Pyrus communis (S)	Pear	1	210	—	—	—	230
Quercus garryana (S)	Garry oak	2	150	—	—	—	650
Quercus marylandica (S)	Maryland oak	56	220	—	45	340	660
Rhamnus purshiana (S)	Cascara	1	140	—	—	—	370
Rhus glabra (S)	Sumach	5	100	—	—	29	570
Ribes lacustre (S)	Swamp gooseberry	1	280	—	—	—	1200
Rosa nutkana (S)	Wild rose	1	280	—	—	18	870
Rubus parviflorus (S)	Thimbleberry	1	240	—	—	—	1200
Salix argophylla (S)	Silverleaf willow	5	200	—	—	21	140
Salix scouleriana (S)	Scouler willow	4	160	—	—	100	1500
Salix sitchensis (S)	Sitka willow	1	260	—	—	—	2800
Sambucus glauca (S)	Elderberry	1	360	—	—	—	560
Shepherdia argentea (S)	Silverberry	1	210	—	—	25	510
Sorbus sitchensis (S)	Serviceberry	2	150	—	—	80	1100
Spirea lucida (S)	Spirea	2	260	—	—	100	2100
Symphoricarpos albus (S)	Waxberry	6	100	—	—	—	640
Taxus brevifolia (S)	Yew	4	160	—	—	—	2000
Thalictrum occidentale (S)	Rue	3	150	—	—	—	4600
Tsuga heterophylla (S)	Western hemlock	7	250	—	—	140	1500
Thuya plicata (S)	Western red cedar	38	120	—	—	56	600
Vaccinium membranaceum (S)	Huckleberry	2	180	—	—	71	840
Vaccinium scoparium (S)	Blueberry	1	210	—	—	—	780
NEW ZEALAND (Brooks and Lyon, 1966; Lyon et al., 1968; Lyon, 1969; Lyon and Brooks, 1969; Nicolas and Brooks, 1969; Timperley et al., 1970a)							
Beilschmiedia tawa (L)	Tawa	35	70	—	—	80	250
(S)	Tawa	35	150	—	—	100	300
Cassinia vauvilliersii (L)	Cottonwood	4	70	—	—	—	500
Hebe odora (L)	Koromiko	4	90	—	30	—	480
Leptospermum scoparium (L)	Manuka	4	63	—	15	—	310
Myrsine salicina (L)	Myrsine	10	75	6	17	25	—
Nothofagus fusca (L)	Red beech	10	90	—	70	—	260
(S)	Red beech	10	70	—	50	—	120
Nothofagus menziesii (L)	Silver beech	10	130	—	35	—	500
(S)	Silver beech	10	80	—	30	—	300
Olearia rani (L)	Tree daisy	20	110	4	12	25	500
(S)	Tree daisy	20	160	8	—	—	500

TABLE 12-4 (Continued)

Species	Common name	No.	Elements (ppm)				
			Cu	Mo	Ni	Pb	Zn
Quintinia acutifolia (L)	Quintinia	10	80	7	30	—	220
(S)		10	150	—	20	—	700
Schefflera digitata (L)	Five fingers	6	110	—	—	40	350
Weinmannia racemosa (L)	Kamahi	10	70	4	27	25	250
(S)		10	140	—	20	—	300
UNITED STATES (Cannon, 1960b)[a]							
Conifers (L)		145	133	5	57	75	1127
Deciduous trees (L)		118	249	7	87	54	2303
Grasses (L)		28	119	34	54	33	850
Other herbs (L)		217	118	19	33	44	666
Shrubs (L)		104	223	15	91	85	1585

L = leaves or needles.
S = stems or twigs.

[a]Copyright © 1960 by the American Society for the Advancement of Science.

also shows an estimate for the same elements in vegetation in general. It is hoped that some of the data in this table will be of assistance to field workers in assessing what levels of elements in particular species are indicative of mineralization. It must be emphasized that information from a table of this nature will be quite insufficient to replace an orientation survey but can nevertheless be helpful in reducing the amount of work that has to be undertaken when such a survey is carried out.

SELECTED REFERENCES
General: Cannon (1960a, 1964); Carlisle and Cleveland (1958); Malyuga (1964); Fortescue and Hornbrook (1967)

CHAPTER THIRTEEN
SIMPLE STATISTICAL ANALYSIS OF BIOGEOCHEMICAL DATA

A sound statistical interpretation of the data is the cornerstone to successful use of the biogeochemical method. Without statistics, the observer can only make qualitative interpretations that are subject to his own unconscious bias and rely largely on instinct and experience. Since different persons will vary greatly in their degree of possession of these two attributes, an assessment of a single set of data by different workers could well result in a wide range of conflicting opinions. The correct sequence of events should be to lay a sound statistical foundation for subsequent interpretation of the data in which the biogeochemist will still be able to use his instinct and experience to the full.

The lack of a sound statistical basis for biogeochemical data has been a disadvantage in earlier work in this field; the reasons are not difficult to discover. It is only in the last decade that computers have become freely available. The computer can complete in a few minutes a series of complicated calculations that would have taken a statistician weeks to complete manually. The labor involved in even a relatively simple calculation, such as the correlation coefficient, renders the operation impracticable for a large volume of data. It is not surprising therefore that earlier workers were not able to apply statistics to the full in their investigations. Despite this disadvantage, there are many examples in the literature where intelligent assessment of the data and the use of less sophisticated semistatistical methods have enabled workers to compensate partially for the lack of computers.

It must not be supposed from the above discussion that a blind overreliance on statistics is recommended, but it is suggested that instinct and experience applied to statistically analyzed data must ultimately be far more effective than the same attributes applied to disorganized raw data.

The level of discussion in this chapter is not intended to be that of an exhaustive treatise on statistics but instead is an attempt to present to the practical field worker a few basic ideas on the statistical treatment of biogeochemical data and to show how the computer may be used to good effect in the evaluation of the information.

13-1   NORMAL AND LOG-NORMAL DISTRIBUTIONS OF THE ELEMENTS

Over a decade ago, Ahrens (1954) published a provocative paper in which he suggested that the concentrations of most chemical elements are *log-normally* distributed in felsic and mafic rocks. His paper initiated a lively controversy which still persists today (Chayes, 1954; Miller and Goldberg, 1955; Aubrey, 1956; Vistelius, 1960). Some of the arguments and counter arguments have been summarized by Shaw (1961).

The normal versus log-normal discussion arises from the common problem in geochemistry and biogeochemistry of obtaining a value for the "average" chemical composition of rocks, soils, or plants. If all the results for a set of data concerning the abundance of a particular element are plotted as a histogram with linear coordinates, a symmetrical or skewed distribution will arise provided that only one population is present. If the histogram is symmetrical, the median, the arithmetic mean, and the mode of the data will coincide at the peak of the histogram. The "average" value mentioned above, therefore, will be represented by either the mean, the median, or the mode.

If the histogram is asymmetric with a pronounced skew, it is possible that a normal distribution will be obtained if the logarithms of the elemental concentrations are used. In such a case, the median of the data will correspond to the geometric mean and the population is said to be log-normally distributed.

Figure 13-1 shows a set of two histograms from a single population plotted on both linear and logarithmic scales. This is a good illustration of a log-normal distribution. Alternatively, a simple calculation can replace the histogram. It is only necessary to calculate the arithmetic and geometric means and compare them with the median. Agreement between the median and geometric mean signifies lognormality; between the median and arithmetic mean, agreement indicates normality. For the data shown in Figure 13-1, the median was 0.70 ppm and the arithmetic and geometric means were 1.2 ppm and 0.95 ppm, respectively. It is clear from this case that the arithmetic mean would have given a value too high to represent a good average value for the population and that the geometric mean would have given a closer approximation.

13-2   MIXED POPULATIONS IN GEOCHEMICAL OR BIOGEOCHEMICAL DATA

The discussion so far has centered around homogeneous populations, but these are somewhat rare in geochemical or biogeochemical data collected in the vicinity of an anomaly. In a mineralized area, there may be two or more different populations of an elemental concentration in rocks, soils, or plants. There will probably be a population representing the normal background values and another representing anomalous levels from an ore body or dispersion halo.

It is common practice for field workers to plot histograms for soil or rock data and then decide what value shall be taken as being anomalous for the area. This procedure is quite straightforward when the two populations are quite separate as shown in Figure 13-2. In Figure 13-2b, however, the two populations are nearly coincident and there would be some difficulty in deciding what concentration levels

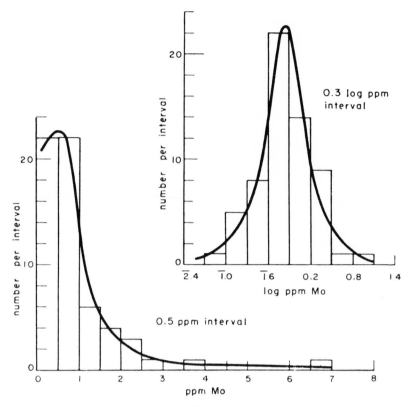

FIGURE 13-1   Molybdenum concentrations in granites illustrating a log-normal distribution.   From Ahrens (1954).   By courtesy of Geochimica et Cosmochimica Acta.

should be taken as representing mineralization and what would represent the threshold values between the anomaly and the background.

The above problem has been discussed by Tennant and White (1959), who have shown that if the cumulative frequency of the data is plotted on probability paper, each population will be represented by a straight line or curve. The intersection of the lines or curves represents a level which may be taken arbitrarily as being the threshold level between background and the anomaly. Whether or not individual populations will appear as straight lines depends on whether their type of distribution (normal or log-normal) is matched with the type of scale (linear or logarithmic) used as the ordinate of the probability paper. Figure 13-3 shows a cumulative frequency diagram of molybdenum concentrations in the ash of *Olearia rani* plotted on logarithmic probability paper. Two separate distributions are evident. If the data had been plotted on arithmetic probability paper, the diagrams would have appeared as curves, but the point of intersection would have been the same.

Threshold levels determined from cumulative frequency diagrams are convenient

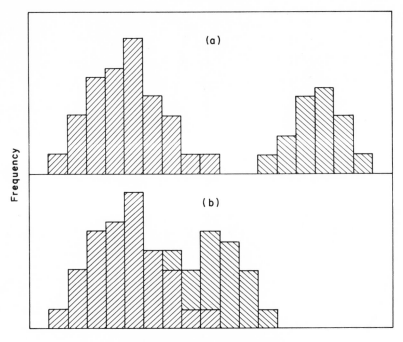

Frequency

Arbitrary concentration units

FIGURE 13-2 (a) Histograms for separate populations representing anomalous and background values for concentrations of an element in soils or vegetation with both distributions well separated. (b) As above but with considerable overlap of the two populations.

values to use in drawing elemental isoconcentration curves on contour maps after biogeochemical orientation or exploration surveys. Figure 13-4 shows a molybdenum anomaly at Takaka in New Zealand outlined by analysis of the soils and by the ash of *Olearia rani*. In each case threshold values from cumulative frequency diagrams were used.

Although two or more distributions are frequently observable in plant material they can be accounted for by reasons other than mineralization in the soil. For example, two background formations may have the same elemental content but the availability of this element to plants may be different. This could produce two distributions in a cumulative frequency diagram, one of which might be assessed incorrectly as an anomalous population. Another explanation of two distributions in plant material is that when the concentration level of element in the soil exceeds a certain threshold value, the plant adopts a different mode of uptake involving perhaps the breakdown of an exclusion mechanism. Although not necessarily related to mineralization, the above threshold level has some use insofar as it indicates an increased elemental concentration in the soil.

The intersection of two cumulative frequency lines or curves does not necessarily indicate a sharp discontinuity between two populations and there can be

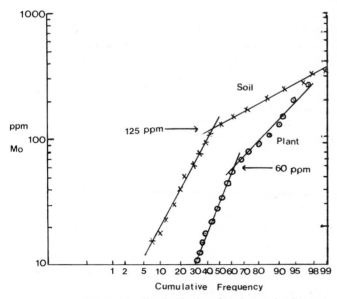

FIGURE 13-3   Threshold values for plant and soil anomalies for molybdenum (Takaka, New Zealand), obtained from a plot of cumulative frequencies on logarithmic probability paper.   From Lyon and Brooks (1969).   By courtesy of the New Zealand D.S.I.R. Information Service.

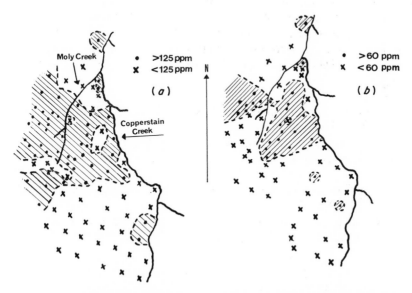

FIGURE 13-4   Molybdenum anomaly at Copperstain Creek, Takaka, New Zealand. (a) Anomaly obtained from soil analyses. (b) Anomaly obtained from analysis of ash of *Olearia rani.* From Lyon and Brooks (1969). By courtesy of the New Zealand D.S.I.R. Information Service.

considerable overlap on each side of the intersection. The two distributions may be separated (Cassie, 1954) and the degree of overlap calculated. Williams (1967) has given worked examples of this method of separation. A more detailed interpretation of cumulative frequency diagrams has been given by Herdan (1948).

The method of plotting data on probability paper is quite simple. The values are arranged in decreasing order of abundance and the individual points are plotted with the cumulative frequency along the x-axis and the concentration values along the y-axis. If there are a large number of points, the data may be grouped and the highest value in each group is recorded.

13-3   SIGNIFICANT DIFFERENCES BETWEEN TWO SETS OF DATA

One of the problems frequently encountered in the interpretation of biogeochemical or geochemical data is whether there is a significant difference between the averages of two sets of values. The field worker may, for example, have a set of analyses for a number of plant samples from a background area and wish to compare these with another set from an unknown area which may or may not be mineralized. The information which is needed is the answer to the following question. *Is there a difference between the means and, if so, with what degree of confidence is the statement made that a difference exists?* It is not sufficient merely to take the means of both sets of data and then assess whether one is higher than the other, since this takes no account of the variability of values within a set.

There are several methods of making the correct statistical calculation to answer the above question. They lead to the well-known Student's t test of significance. The layout of the following worked example is based on Hinchen (1969).

As a result of a field survey, the following concentrations of copper were obtained in the ash of a species of vegetation sampled from two different areas. It is required that it be established whether or not the two sets have statistically different means.

> DATA FROM SET 1 (*ppm copper*): 25.4, 26.1, 23.7, 20.9, 18.4, 27.1, 30.9, 21.1, 23.4, and 22.4.
>
> DATA FROM SET 2 (*ppm copper*): 27.1, 26.3, 28.9, 27.7, 29.3, 20.7, 21.6, 28.9, 29.8, and 30.1.

1. Calculate the respective means ($\bar{x}_1$ and $\bar{x}_2$). These are 23.94 and 27.04, respectively.

2. Calculate the standard deviations (S_1 and S_2) for each set from the formula:

$$S = \left(\frac{\Sigma d^2}{n-1} \right)^{\frac{1}{2}}$$

where n is the number of samples in a set and d is the deviation of each sample value from the mean. The standard deviations are 3.58 and 3.33, respectively.

3. Pool the standard deviations from the formula:

$$S_p = \left[\frac{(n_1 - 1)S_1{}^2 + (n_2 - 1)S_2{}^2}{n_1 + n_2 - 2}\right]^{\frac{1}{2}} = \left[\frac{(9 \times 3.58^2) + (9 \times 3.33^2)}{18}\right]^{\frac{1}{2}}$$

$$= 3.46$$

4. Calculate Student's t for the comparison of two means, using the formula:

$$t = \frac{\overline{x}_1 - \overline{x}_2}{S_p\left(\frac{1}{n_1} + \frac{1}{n_2}\right)^{\frac{1}{2}}} = \frac{-3.10}{3.46\sqrt{0.2}} = -1.99$$

5. The term $n_1 + n_2 - 2$ is the number of *degrees of freedom*.

Now select an arbitrary confidence limit such as 95% and obtain the t value from Appendix I for 18 degrees of freedom and for the next highest t value above 95% (i.e., $t_{0.975}$); this gives a value of ± 2.10 for t. This must be compared with the experimental value of -1.99 which lies inside the range of values for t at the 95% confidence limit. There is clearly no significant difference between the two sets of biogeochemical data at this level. If the standards are now lowered slightly to the 90% level, the value for $t_{0.95}$ is ± 1.73 and since the experimental value of -1.99 lies outside these limits, there is clearly a difference in the original two sets of data at the 90% confidence limit.

The results of the t test should not be accepted uncritically. It will be noted from Appendix I that it will be possible to show a significant difference between almost any pair of different sets of data provided that enough samples are taken. For example, in the above case, $t_{0.975}$ would have been ± 1.98 for 120 degrees of freedom (61 samples) and there would then have been a just-significant difference between the two sets of data at the 95% confidence limit. There is a finite limit to the improvement in confidence limits with increasing numbers of samples. It is clear from Appendix I that, for an infinite number of samples, Student's t has a limiting value of ± 1.96 at the 95% confidence level so that in the above example, if the experimental value of t had been -1.94 instead of -1.99, it would never have been possible to show a statistical difference between the means at the 95% confidence level.

13-4   THE RELATIONSHIP BETWEEN TWO SETS OF DATA

The main requirement of a biogeochemical orientation survey is to establish what relationship exists between the concentration of an element in a plant organ and its concentration in the soil or bedrock. If the relationship is very obvious, it will probably be sufficient to plot the points on a graph and visually draw a line or curve through them. In practice, relationships are seldom so simple and sometimes the decision as to whether there is a relationship depends largely on individual opinions. If a number of plants species have been tested, several questions have to be answered: *Is there a relationship between elemental concentrations in plants and soil? If this is so, what species or organs should be sampled?*

Other questions involve threshold levels in plants indicative of anomalies in the soil or rock and the problem of whether results should be expressed on an ash-weight or dry-weight basis.

REGRESSION ANALYSIS   When scattered points are plotted on a graph, the main problem is to find the line of best fit. This term is vague and, if a linear relationship is assumed, a more definite quantity is the *least squares line* which is one for which the sum of the squares of the coordinate distances of each point from the line is a minimum.

The assumption is made that the x-variable (abscissa) is independent and not subject to error, whereas the y-variable (ordinate) is dependent. The term "regression of y on x" implies that x is the independent variable whereas the reverse, the "regression of x on y" implies the opposite. The regression lines are different in each case.

Assuming that $y = f(x)$ and consequently $y = mx + c$ for a linear relationship, the problem in finding the regression line is to evaluate the constants m and c.

It can be shown that:

$$m = \frac{n \Sigma xy - \Sigma x \Sigma y}{n \Sigma x^2 - (\Sigma x)^2} \tag{13.1}$$

and

$$c = \bar{y} - m\bar{x} \tag{13.2}$$

where the terms $\bar{x}$ and $\bar{y}$ have the same meanings as in Section 13-3.

As a worked example it may be shown how regression analysis can be used to calculate the regression line for a set of biogeochemical data representing the nickel content of plant ash expressed as a function of the content of this element in the soils. The data are shown in Table 13-1.

If values from the table are substituted in equations (13.1) and (13.2), the following results are obtained.

$$m = \frac{(10 \times 24.725) - (455 \times 475)}{(10 \times 24.575) - (455)^2} = 0.80$$

TABLE 13-1   Calculation of the Regression Line for a Plot of Nickel Concentrations (ppm) in Plant Ash (y) as a Function of the Nickel Content (ppm) of the Soil (x)

x	y	x^2	xy
10	15	100	150
20	30	400	600
30	45	900	1,350
40	30	1,600	1,200
45	45	2,025	2,025
50	55	2,500	2,750
55	65	3,025	3,575
65	45	4,225	2,925
70	70	4,900	4,900
70	75	4,900	5,250
$\Sigma x = 455$	$\Sigma y = 475$	$\Sigma x^2 = 24,575$	$\Sigma xy = 24,725$

and

$$c = 47.5 - (0.80 \times 45.5) = 11.1$$

Hence

$$y = 0.80x + 11.1$$

This line (the regression of y on x) is shown in Figure 13-5 which is a plot of the data of Table 13-1.

To obtain the regression of x on y, it is only necessary to interchange values of x and y in Table 13-1 and recalculate as before. This gives the following equation:

$$x = 0.91y + 0.77$$

This line is also shown in Figure 13-5.

The two regression lines are seldom the lines which would have been drawn by visual methods. There is, moreover, a fallacy in the use of regression lines for plant–soil data. The assumption is made that x is an independent variable free of

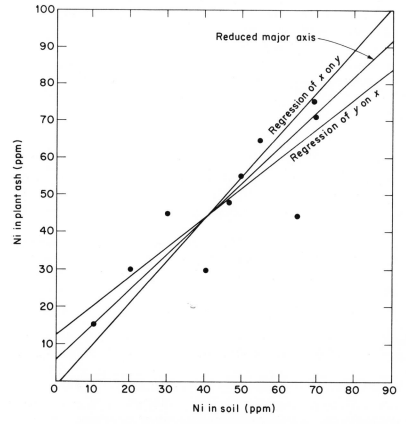

FIGURE 13-5 Regression lines and reduced major axis for plot of nickel concentrations in plant ash (y-axis) as a function of the nickel content of the soil (x-axis).

error. If x represents elemental concentrations in soils there is no reason to suppose that soil values are any more free of error than those for vegetation. Soil data are subject to analytical error as well as many others.

The problem of this deficiency in regression analysis applied to geologic problems has been discussed by Imbrie (1956) who has recommended that use should be made of the so-called *reduced major axis* in placing a line through a set of points. This, according to Imbrie "(i) makes no assumption of independence; (ii) is invariant under change of scale; (iii) is simple to compute; and (iv) produces results intuitively more reasonable than corresponding results obtained using regression analysis." Figure 13-5 also shows the reduced major axis which lies between the two regression lines and is probably closer to the line that would have been drawn by visual means alone.

CORRELATION ANALYSIS   In the previous subsection, it was shown how a regression line can be fitted to a series of points on a graph. One vital question remains unanswered however. *How closely do the points cluster around the line?* In other words, *how good is the linear relationship at predicting y from x?* This is an important question in a biogeochemical orientation survey where concentrations of an element in plants have to be compared with geophysical data or with corresponding values in the soils or bedrock.

A good approach to the problem is the use of the *Pearson product moment correlation coefficient (r)*. This may be expressed mathematically as follows:

$$r = \frac{\Sigma xy - \left(\dfrac{\Sigma x \Sigma y}{n}\right)}{\left[\left(\Sigma x^2 - \dfrac{(\Sigma x)^2}{n}\right)\left(\Sigma y^2 - \dfrac{(\Sigma y)^2}{n}\right)\right]^{\frac{1}{2}}} \tag{13.3}$$

The following is a worked example for a calculation of r from the data in Table 13-1. The number of samples is far fewer than would normally be employed but has been restricted for space considerations and ease in calculation. The sequence of calculation (based on Hinchen, 1969) is as follows.

Calculate Σx (=455), Σy (=475), Σx^2 (=24,575), Σy^2 (=25,875), Σxy (=24,725), $(\Sigma x)^2/n$ (=20,700), $(\Sigma y)^2/n$ (=22,560), and $(\Sigma x \Sigma y)/n$ (= 21,610). Substituting these values in equation (13.3) we obtain the relationship:

$$r = \frac{24,725 - 21,610}{[(24,575 - 20,700)(25,875 - 22,560)]^{\frac{1}{2}}} = 0.87$$

The correlation coefficient can vary between the limits of ± 1. Positive values of r indicate that the two variables are directly related; negative values imply an inverse relationship. Values approaching $+1$ or -1 indicate increasing linearity between the two variables, whereas 0 implies no linear relationship whatsoever.

In the above calculation, a value of 0.87 was obtained for r and implies that a direct relationship exists. To establish the confidence which may be put on the relationship, the probability value should be obtained from Appendix II for $n - 2$ degrees of freedom. For 8 degrees of freedom as in the above example, the value lies just over 0.001. This value is the probability that a relationship does *not* exist and implies that there is almost a 99.9% chance that there *is* a linear relationship between the two sets of data and that it is not due to random chance. Examination of Appendix II will show that the significance of *r* is a function of the number of degrees of freedom and that it is similar to Student's *t* in this respect.

It is customary to divide significance into a number of divisions. Brookes et al. (1966) have suggested the following scheme of classification for various values of the probability function: Probability greater than 0.1, not significant (NS); between 0.10 and 0.05, possibly significant (PS); between 0.05 and 0.01, significant (S); between 0.01 and 0.001, highly significant (S*); probability less than 0.001, very highly significant (S**).

It must be emphasized that these classifications are somewhat arbitrary and under certain circumstances even a 90% chance that there is a relationship (probability 0.10) may be acceptable though it is usually considered to be not significant.

Correlation coefficients and levels of significance are invaluable in biogeochemical orientation surveys. This is illustrated in Table 13-2. The table gives a large amount of information. It is clear in this example that the best vegetation for determining zinc in soils would have been the leaves of *Schefflera digitata* expressed on a dry-weight basis; whereas for lead, the best results would have been obtained with the wood ash of *Beilschmiedia tawa*. Neither species was suitable for copper.

The correlation coefficient as expressed in Eq. (13.3) provides a valid probability estimate only if the data are normally distributed. If either set of variables has a log-normal distribution, it should be transformed logarithmically to preserve the validity of the probability estimate.

There are clearly four combinations of two variables *A* and *B* with the option of two types of scale for each. That is, linear *A* vs. linear *B*, linear *A* vs. logarithmic *B*, logarithmic *A* vs. linear *B*, logarithmic *A* vs. logarithmic *B*. In order to make the correct decision in this matter, plot cumulative frequency diagrams or histograms for each variable to see how the elemental concentrations are distributed. Alternatively, test each variable as in section 13-1 for normality or log-normality.

Conclusions based on correlation coefficients from nonnormal data should not be given much weight unless the correct logarithmic transformations have been performed. One common example is a graph exhibiting a small number of points

TABLE 13-2   Correlation Coefficients for the Logarithmic Relationship Between Copper, Lead, and Zinc in Soils and in Vegetation at Te Aroha, New Zealand

				Copper		Lead		Zinc	
Species	Organ	State	No. of samples	*r*	Significance	*r*	Significance	*r*	Significance
Schefflera digitata	Leaves	Dried	36	0.05	NS	0.27	NS	0.45	S*
Schefflera digitata	Leaves	Ashed	36	−0.03	NS	0.28	NS	0.41	S
Beilschmiedia tawa	Leaves	Dried	35	−0.05	NS	0.26	NS	−0.04	NS
Beilschmiedia tawa	Leaves	Ashed	35	−0.03	NS	0.31	PS	0.06	NS
Beilschmiedia tawa	Wood	Dried	28	0.26	NS	0.45	S*	0.22	NS
Beilschmiedia tawa	Wood	Ashed	28	0.24	NS	0.48	S*	0.29	NS

SOURCE: Nicolas and Brooks (1969).

TABLE 13-3   Correlation Coefficients (r) for the Relationship Between the Concentration of Molybdenum (ppm) in the Ash of *Olearia rani* and in the Soil at 45 Sampling Sites

Scale for expressing concentrations in plants	Scale for expressing concentrations in soils	r	Significance
Linear	Linear	0.47	S*
Linear	Logarithmic	0.48	S*
Logarithmic	Linear	0.70	S**
Logarithmic	Logarithmic	0.77	S**

with very much higher values than the majority. For example, Figure 13-6a shows a graph of copper values in a plant species expressed as a function of the copper content of the soil. It will be noted that there are a number of closely spaced points clustered near the origin and two separate points with very high values. Statistically, the graph will virtually behave as three points, with the cluster behaving as a heavily weighted single point. The value for r is hence quite high ($r = 0.92$). If the two high points had been ignored and the remaining 30 values expanded to a wider scale, a graph similar to that of Figure 13-6b would have been produced. The value of r for the remaining points is only 0.26, which is not significant. To summarize, the value of r obtained by statistical analysis can only be meaningful if the data are normally distributed or are logarithmic transformations from a log-normal distribution. This factor should always be considered carefully in assessing correlations.

It has been mentioned above that the Pearson correlation coefficient suffers from the disadvantage that it is not valid if the data are not normally distributed. This problem can be avoided if, instead, the Spearman rank correlation coefficient is used. This coefficient is nonparametric and can best be explained as follows:

If there are a number of pairs of data for which correlation has to be determined, the results are arranged in two columns. The first column is ranked with the highest values at the top and the lowest values at the bottom (or vice versa). The values in the second column are then arranged in exactly the same order as those of the first column so that all pairs are matched horizontally. If there is a complete correlation, then the ranks of the second column will be exactly as in the first and the Spearman coefficient (r_s) will be 1.0. The extent of deviation from perfect ranking in the second column will determine the value of r_s which ranges from -1 to $+1$.

In its simplest form r_s can be represented as follows:

$$r_s = 1 - \frac{6 \, \Sigma \, d^2}{n^3 - n} \tag{13.4}$$

where n is the number of pairs of values and d is the difference of rank of each horizontal pair in the two columns. When there are tied values in either column, this formula is no longer valid and a modification has to be used. This is:

$$r_s = \frac{A + B - \Sigma \, d^2}{2(AB)^{\frac{1}{2}}} \tag{13.5}$$

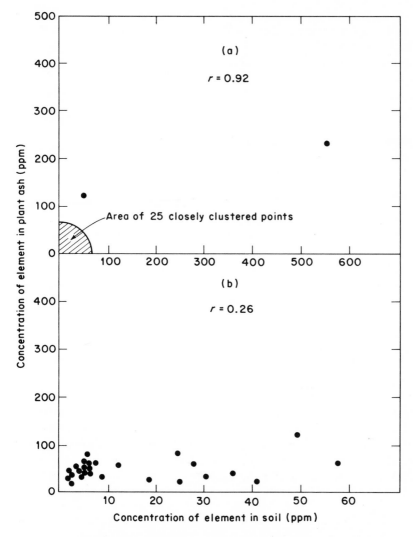

FIGURE 13-6 Hypothetical values for concentrations of an element in plant ash and in soils.

Plot (a) includes two abnormally high values and gives a highly significant value for the correlation coefficient (r) between the two variables.

Plot (b) ignores the two high values and shows that the resultant value of r implies a nonsignificant relationship.

where

$$A = \frac{n^3 - n}{12} - \Sigma T_a \quad \text{and} \quad B = \frac{n^3 - n}{12} - \Sigma T_b$$

T_a and T_b are corrections for the presence of ties and have each the value $(t^3 - t)/12$, and t is the number of observed ties in column a or column b.

The following is a worked example using the data from Table 13-1. The value of r_s hence obtained with these data will be directly comparable with the value of r obtained earlier.

The data in the first column of Table 13-1 are ranked and the results are: 1, 2, 3, 4, 5, 6, 7, 8, 9.5, 9.5. The last two values represent a tie and in such cases it is necessary to assign to each the means of ranks which would have been assigned if no tie had occurred. The ranking of the second column as a consequence of rearrangement of the first, then gives values of: 1, 2.5, 5, 2.5, 5, 7, 8, 5, 9, 10. Note that in this column there is one tie of two values and one of three. Hence,

$$A = \frac{10^3 - 10}{12} - \frac{8 - 2}{12} = 82$$

and

$$B = \frac{10^3 - 10}{12} - \frac{8 - 2}{12} - \frac{27 - 3}{12} = 80$$

The rank differences for each pair taken horizontally across the columns are: 0, −0.5, −2, 1.5, 0, −1, −1, 3, 0.5, −0.5 From this

$$\Sigma d^2 = 18$$

Hence,

$$r_s = \frac{82 + 80 - 18}{2(82 \times 80)^{\frac{1}{2}}} = 0.89$$

The above value compares with 0.92 for the Pearson coefficient. The significance of r_s is, however, only 91% that of r. That is, if the table in Appendix II is used to obtain significance values, and if, for example, 100 degrees of freedom were involved, a value of 0.321 for r should be equivalent to 0.337 for r_s (91 degrees of freedom). Appendix II can therefore be used for rank correlation coefficients by reducing the number of degrees of freedom to 91% whenever the significance of r_s has to be obtained.

Although the Spearman rank correlation coefficient does not depend on the type of distribution of the data, this advantage is somewhat offset by the fact that considerably more computer time is required for making this calculation because of the time involved in carrying out the ranking part of the program. In spite of this problem, the rank coefficient is in general more satisfactory than the Pearson coefficient particularly where a small number of samples are involved and where it is impossible to decide whether the data are normally or log-normally distributed.

13-5   THE COMPUTER

As has already been mentioned, the wide availability of computers in recent years has much facilitated the science of statistics and now it is feasible to carry out correlations for a large number of samples and for many pairs of sets of data. A computer program written in FORTRAN II–D for use with an IBM 1620 II computer has been prepared at Massey University for the treatment of biogeochemical data. This program, which is an extensive modification by M. H. Timperley of an original by Lyon (1969), is used to calculate means, standard deviations, Pearson correlation coefficients, and gradients of reduced major axes (see Appendix III). The program has been written in a simple form to enable prospective users to follow readily the computational steps. A program for calculation of rank correlation coefficients is shown in Appendix IV.

For application to biogeochemical data, it was convenient to use one card for each sample site; thus the first card would contain up to 12 variables concerning the sample at site one (e.g., 12 elemental concentrations). This system enables all the data at any particular sample site to be readily removed from the computations if desired. It also facilitates data print-out.

The data must be punched on to cards in 12 columns each 6 spaces wide: column 1 occupies positions 1–6, column 2 occupies columns 7–12, and so on up to column 72. Columns 73–78 are used for the sample reference label (e.g. sampling position). This column is not included in the computation but is printed out with the data.

In each column of 6 spaces, the first space from the left must be left vacant; and if the number to be punched has a decimal point, this must be punched also. In other words, a number containing a decimal point will have at the most 4 significant figures on the card. A title card with the numeral 1 in column 1 and the title in the next 49 spaces is placed immediately before the data cards and a card with the numeral 9 in column 80 is placed immediately behind the data cards.

The majority of biogeochemical data considered at Massey University tend to be log-normally distributed, and for computational purposes the program transforms all input data to base ten logarithmic form before carrying out the calculations.

The geometric means and standard deviations for each of the 12 columns are printed out, followed by the 66 correlation coefficients and gradients of the reduced major axes. The latter two sets of results are printed out in matrices of the columns.

Two optional transformations of the input data are possible according to the configurations of sense switches 1 and 2. Examples of the use of these options are included in the program. The first is the multiplication of columns 1 to 10 by column 11. This could be used to convert dry-weight values to ash weights by multiplying by the percentage ash weight in column 11. The second is the division of each odd-numbered column by the next even-numbered column with the placement of the result in the odd-numbered column. This could be used to obtain elemental ratios in plants or soils.

The emphasis in this chapter has been on simplicity. There are other statistical methods such as *trend surface analysis* and *multiple regression* which might be used to advantage in biogeochemical prospecting. A discussion of more advanced statistical methods is outside the scope of this book and the reader is referred to works such as Griffiths (1967) for a fuller treatment of statistics applied to geologic problems.

SELECTED REFERENCES
Simple statistics: Hinchen (1969)
Statistics in geology: Griffiths (1967); Middleton (1963)

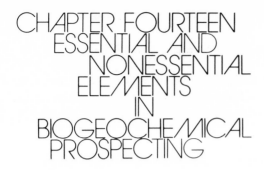

CHAPTER FOURTEEN
ESSENTIAL AND
NONESSENTIAL
ELEMENTS
IN
BIOGEOCHEMICAL
PROSPECTING

The comparatively extensive literature on biogeochemical prospecting during the past three decades contains several references to failure as well as success. As knowledge of the technique has accumulated, some elements have achieved the reputation of being "easy" for prospecting while others are said to be "difficult." Elements such as molybdenum and uranium belong to the first group, whereas copper and zinc appear to have caused difficulty in the past. The position was summarized by Marmo (1953). During work carried out in Finland he observed that:

> From observations made by botanists, the concentration of copper in plants does not increase linearly but after reaching some value separately determined for each plant, the increase of the copper content of the plant will no longer reflect the increasing copper content of the underlying rock as well as when comparatively small copper contents are in question.

A similar failure of plants to indicate zinc has been reported by several workers. For example Boyle (1958) observed a poor response of vegetation to a zinc anomaly in the Yukon region of Canada.

If copper and zinc are "difficult" elements, this fact is masked somewhat by the preponderance of literature concerning these two elements so that even if the failure rate is higher than with some other elements, successful applications of the method will still far outnumber those for all other elements combined.

After several years of research in New Zealand, the author and his co-workers have carried out several biogeochemical investigations (Brooks and Lyon, 1966; Lyon et al., 1968; Lyon and Brooks, 1969; Nicolas and Brooks, 1969; Whitehead and Brooks 1969b; Timperley et al., 1970a), but in all cases, results were disappointing for copper and zinc. Statistical analysis of all available data was subsequently undertaken with a view to ascertaining whether there were any peculiar features of accumulation of these two elements by plants which were not applicable to other elements.

14-1   STATISTICAL ANALYSIS OF BIOGEOCHEMICAL
DATA FOR COPPER AND ZINC

One of the regular tests used by the author's research group for the statistical analysis of biogeochemical data involves a computer program (see Chapter 13) in which, inter alia, the logarithms of relative uptakes (enrichment coefficients) of elements by plants are correlated against the logarithms of the concentrations of the elements in the soils. In the course of this work, it was noted that values of the correlation coefficients (r) indicated a highly significant inverse relationship between these two variables for copper and zinc in several plant species. This inverse relationship was very obvious visually when plotting the data on logarithmic paper.

The above highly significant correlations contrasted sharply with the non-significant correlations obtained for the direct plant–soil relationships of the concentration of an element in plant ash as a function of the amount in the soil (logarithmic coordinates).

When linear coordinates were used for the plots of relative uptake as a function of the soil content, a very good approximation to a rectangular hyperbola was obtained. This relationship was observed for copper and zinc in all of seven species of plants. Figure 14-1 shows linear and logarithmic plots for copper in *Quintinia acutifolia*. Figure 14-2 shows the same plots for zinc in the same species. Both figures are from Timperley et al. (1970b), who originally reported this anomalous behavior.

Although hyperbolic plots were obtained for every case involving the biogeochemistry of copper and zinc in New Zealand, none of the other elements investigated (chromium, cobalt, lead, molybdenum, nickel, and uranium) showed a similar behavior. This is demonstrated in Figure 14-3 which shows (linear coordinates) enrichment coefficients for copper, molybdenum, and nickel in three species of New Zealand plants, plotted as a function of the soil content of each element. It will be noted that the data for molybdenum and nickel lie along a straight line parallel to the *x*-axis.

14-2   THE SIGNIFICANCE OF ESSENTIAL ELEMENTS
IN BIOGEOCHEMICAL PROSPECTING

The hyperbolic plots in Figure 14-3 indicate that the relative uptake of copper and zinc decreases sharply as the concentration of these elements in the soil increases, and reaches a definite threshold level at which point relative uptake still decreases but at a much slower rate.

It is believed that this mechanism is linked to the fact that copper and zinc are essential elements for plant growth. The concentration in the plant at the threshold level probably represents the upper limit of the plant's physiological requirement concentration for the element concerned. When the content in the soil increases beyond this level, the plant appears to operate a partial rejection of the soil element.

It is clear from the above discussion that hyperbolic plots similar to the above may be an indication that an element is essential to the plant concerned. This type of plot is, however, not evident for molybdenum which is known to be an essential

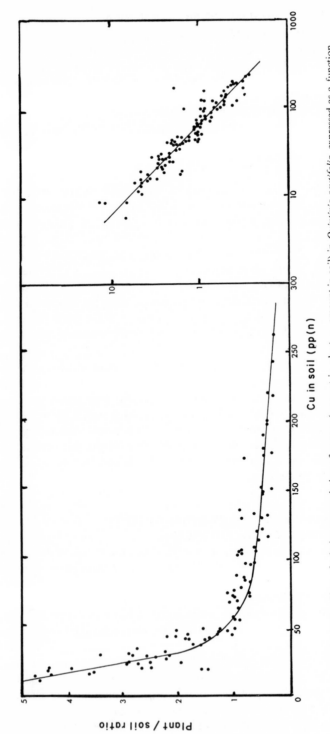

FIGURE 14-1 Plots of relative accumulation of copper (amount in plant ÷ amount in soil) in *Quintinia acutifolia*, expressed as a function of the concentration of copper in the soil. Data expressed in both linear and logarithmic units. After Timperley et al. (1970b). By courtesy of the Journal of Applied Ecology.

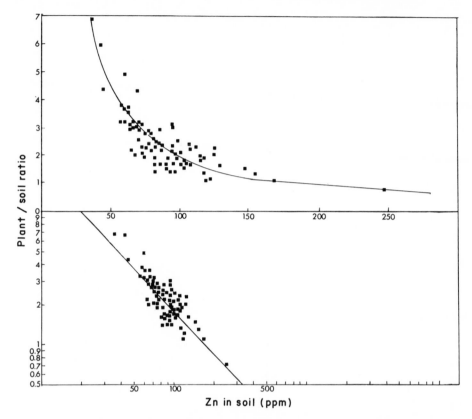

FIGURE 14-2   Plots of relative accumulation of zinc (amount in plant ÷ amount in soil) in *Quintinia acutifolia* as a function of the concentration of zinc in the soil. From Timperley et al. (1970b). By courtesy of the Journal of Applied Ecology.

element. It is probable in this case that the physiological requirement by plants is extremely low and that consequently amounts of molybdenum in the soil never become low enough for the hyperbolic plot to become evident.

Since a perfectly linear plot of the data for essential elements is obtained when logarithmic coordinates are used, there must be a simple linear relationship between the relative uptake and soil content variables. Linearity of the relationship implies that:

$$\log y = n \log x + \log k$$

where y is the relative uptake, x is the concentration of the element in the soil, and k is a constant. Plots of $\log x$ vs. $\log y$, shown in Figures 14-1 and 14-2, give slopes very close to -1. This was also observed for four other examples investigated.

Since $n = -1$,

$$\log y = -\log x + \log k \quad \text{and} \quad \log xy = \log k$$

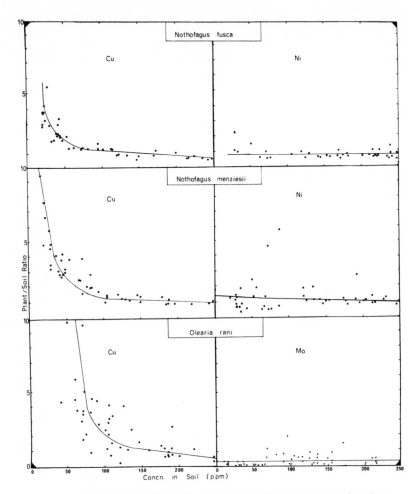

FIGURE 14-3 Plots of relative accumulation (amount in plant ÷ amount in soil) as a function of the soil content for various elements and plant species. From Timperley et al. (1970b). By courtesy of the Journal of Applied Ecology.

Finally, $xy = k$, or in other words the relative uptake will decrease as the content in the soil increases in order to preserve the value of k, which clearly is the plant's physiological requirement level for the element concerned.

When elements are nonessential, enrichment coefficients are apparently constant throughout a wide range of the elemental content in the soil. In other words, there is a linear relationship between the amounts in the plant and soil. This is apparent for the nickel data in Figure 14-3.

For essential elements, the decrease of enrichment coefficients with increasing soil contents of the appropriate element is a very important factor affecting the validity of biogeochemical methods of prospecting. In the part of the hyperbolic plot approaching parallelism with the x-axis, the effect of essentiality may not be very serious but at lower elemental contents in the soil, the biogeochemical method will be ineffective. To consider a specific case, the leaf ash of Q. *acutifolia* (Figure 14-1) would be unsuitable in prospecting for copper in an area containing less than 100 ppm of this element in the soil. The same species (Figure 14-2) would be equally ineffective in prospecting for zinc in soils containing less than 150 ppm of this element.

The author is reluctant to assume that the above relationship is a characteristic of all plants since his own experience is confined to seven New Zealand and one Australian species (*Acacia aneura*), but there can be little doubt that there is probably a general tendency for most plants to keep their contents of essential elements to a constant level regardless of the amount in the soil.

In the course of further investigations into this problem, the author and his co-workers have developed simple statistical and graphical techniques which will enable the field worker to predict the results of an orientation survey without carrying out extensive statistical studies. These simple tests are recorded in the remainder of this chapter and are centered around the question of essentiality or nonessentiality of elements in plants.

14-3   CRITERIA FOR ESTABLISHING ESSENTIALITY OR
        NONESSENTIALITY OF ELEMENTS IN PLANTS
        NORMAL AND LOG-NORMAL DISTRIBUTIONS   There has already been some discussion of normal and log-normal distributions (see Chapter 13). Liebscher and Smith (1968) have suggested that, in human tissue at least, essential elements are normally distributed and nonessential elements are log-normal. A simple statistical test is to compare geometric and arithmetic means with the median. Agreement of the median with the geometric mean implies log-normality; agreement with the arithmetic mean implies a normal distribution. This test, however, is not very satisfactory as it is only valid if data are taken from one distribution only. The presence of anomalous distributions in addition to background populations will nearly always give a skew to the histogram, and the combined data will appear to be log-normally distributed. Table 14-1 shows statistical data for plants and soils. Application of the above test to the data shows that almost without exception, distributions in plants are apparently log-normal. Therefore, it does not appear that Liebscher's test is really applicable to plants or soils. At best it is useful only where single populations are involved.

        PLOTTING RELATIVE UPTAKE AS FUNCTION OF THE SOIL CONTENT OF THE ELEMENT   A quick and approximate method of establishing the essentiality or otherwise of an element in plant material is to plot the relative uptake as a function of the soil content of the element on linear or logarithmic coordinates as in Figures 14-1 and 14-2. The shape of the plot, linear or hyperbolic, will indicate whether the element is essential or not, as has already been mentioned.

TABLE 14-1   Statistical Data for Trace Elements in Plants and Soils

Species	Element	No.	Concentrations in plant ash (ppm)				Concentrations in soils (ppm)				Ratio of cv	r for plant–soil	Whether significance of r could have been predicted from values of cv
			Median	Arithmetic mean	cv (%)	Geometric mean	Median	Arithmetic mean	cv (%)	Geometric mean			
B. tawa	Cu	35	100	132	137	100	60	190	253(L)	62	0.54	−0.03(NS)	Yes
	Pb	35	100	229	157	123	148	963	248(L)	229	0.63	0.31(PS)	No
	Zn	35	270	304	39	282	150	287	169(L)	166	0.23	0.06(NS)	Yes
N. fusca	Cu	49	100	111	75	99	95	148	126(L)	86	0.59	0.62(S**)	No
	Ni	49	90	163	91	124	205	203	49(S)	172	1.85	0.82(S**)	Yes
	U	26	5.0	5.6	100	3	12	20	121(L)	8.9	0.82	0.62(S**)	Yes
	Zn	49	250	263	29	253	81	83	27(S)	81	1.05	0.03(NS)	Yes
N. menziesii	Cu	48	160	160	27	155	75	93	73(L)	71	0.37	0.25(NS)	Yes
	Zn	48	585	602	29	580	83	88	41(S)	82	0.70	0.09(NS)	Yes
	Ni	48	157	193	70	146	152	159	53(S)	131	1.30	0.64(S**)	Yes
O. rani	Cu	71	155	209	61	166	105	149	54(S)	132	1.13	0.08(NS)	Yes
	Mo	71	30	87	252	28	100	143	65(S)	118	3.87	0.74(S**)	Yes
Q. acutifolia	Cu	93	72	77	47	72	80	120	125(L)	80	0.38	0.33(S*)	No
	Ni	93	87	92	43	84	175	181	45(S)	157	0.96	0.57(S**)	Yes
	U	22	3.0	7.8	177	1.4	10	97	271(L)	14	0.65	0.74(S**)	Yes
	Zn	93	200	196	21	192	83	89	31(S)	86	0.68	−0.08(NS)	Yes
S. digitata	Cu	36	130	179	62	138	60	113	192(L)	62	0.32	−0.03(NS)	Yes
	Pb	36	120	174	78	123	148	643	207(L)	229	0.38	0.28(NS)	No
	Zn	36	440	588	71	501	150	222	135(L)	166	0.53	0.41(S)	Yes
W. racemosa	Cu	103	70	77	64	71	80	125	119(L)	81	0.54	0.20(S)	Yes
	Ni	103	105	126	69	102	175	182	47(S)	156	1.47	0.62(S**)	Yes
	U	20	2.0	3.3	130	1.2	6	11	129(L)	5.6	1.00	0.72(S**)	Yes
	Zn	103	220	231	31	221	83	88	32(S)	84	0.97	0.00(NS)	Yes

L = "Large" range.
S = "Small" range.

SOURCE: Timperley et al. (1970b).

THE RANGE OF CONCENTRATIONS OF ELEMENTS IN PLANTS   It might be expected that the concentration range of essential elements in plants would be narrower than that for the nonessential elements because of regulatory mechanisms which tend to keep the amounts of essential elements relatively constant in plants. Table 14-1 gives the coefficients of variation (*cv*) for elemental concentrations in plants and soils from New Zealand and Australia. The coefficient of variation is the standard deviation divided by the mean. It will be noted that values of *cv* for copper and zinc in plants tend to be significantly less than for most other elements. The value of *cv* in plants is therefore in itself some indication of the degree of essentiality of the element concerned. It will also be noted from the correlation coefficients for elements in plants and soils, that values of *r* for copper and zinc tend to be less significant than for other elements but become more significant for elements with wider ranges of concentrations in plant material (greater values of *cv*).

It would seem from the above discussion and from inspection of Table 14-1 that there is some justification for assuming that the magnitude of *cv* values for elements in plants is a criterion of essentiality. This argument does not take into account the range of values of the element in the soil. Although the concentration of an essential element in plant material will be somewhat independent of the soil values, this will not apply at relatively high concentrations in the soil because, under these conditions, the plant's regulatory mechanism will tend to break down. Indeed it is only because of this breakdown that biogeochemical prospecting is possible for "difficult" elements such as copper and zinc.

Timperley et al. (1970b) have suggested that a more meaningful interpretation of essentiality is to take the ratio of the plant and soil values for *cv* (plant/soil). These ratios are shown in Table 14-1.

If the range of values for an element in the soil is small, an equally small variation for the same element should appear in plants whether the element is essential or not. It is only when the range of values in the soil is relatively great, that large variations of *cv* between essential and nonessential elements might be expected in the plant. For this reason, Timperley et al. (1970b) have classified the plant/soil *cv* ratios into two groups depending on the concentration of the element in the soil.

When *cv* for an element in the soil is greater than 70%, this is taken arbitrarily as being a "large" range. The term "small" range is assigned to values less than 70%.

Inspection of Table 14-1 shows that within a species, when *cv* is large for a particular element, the plant/soil ratio of the coefficient of variation is (with the exception of lead in *Schefflera digitata*) always greater for nonessential than for essential elements. If for the above case, a value of 0.60 is taken as a dividing line for ratios of *cv*, nonessential elements are involved for values higher than this and essential elements for those below. A particularly significant observation is that values of *r* for the plant–soil relationship are nearly always highly significant for these nonessential elements and not significant for the essential elements. It would therefore seem that it is possible to predict a favorable or unfavorable plant–soil relationship merely by considering ratios of the coefficients of variation for the concentration of an element in plants and soils.

When the range of values of an element in soils is small, values of *cv* for non-essential elements in plants should decrease and less differentiation would be expected between essential and nonessential elements. It is nevertheless still possible to distinguish statistically between the two classes of elements under these circumstances.

When there is a small concentration range of an element in the soils (*cv* less than 70%), a threshold value of 1.2 has been assigned by Timperley et al. (1970b) for distinguishing between essential and nonessential elements in plants as measured by the plant/soil *cv* ratio. Using this criterion, Table 14-1 shows that it would have been possible to predict essentiality or nonessentiality in every case except for nickel in *Q. acutifolia* and molybdenum in *Olearia rani*. Molybdenum is a special case; and it has already been mentioned that this element behaves statistically as a nonessential element, probably because most plants may have a very low physiological requirement for it.

Overall, essentiality or nonessentiality would have been predicted correctly in 21 out of 23 cases using the above criteria. In a further 19 out of 23 cases, it would have also been possible to predict accurately whether or not a highly significant correlation would exist between an elemental content in plants and in soils. A highly significant correlation is taken as having a significance at the 99% confidence level or greater (i.e., S*; see Chapter 13).

The above statistical tests are summarized in Table 14-2.

As an illustration of the above discussion, two specific examples from Table 14-1 may be considered. Uranium was determined in the leaf ash of *Q. acutifolia*. The coefficients of variation for the concentration of this element in plant material and soils were 177% and 271%, respectively. The value for the soil implied a large range for which the threshold level of the ratio of the *cv* values is 0.60. The observed value of 0.65 indicated that uranium in the plant was nonessential and that *r* for the plant–soil relationship would be very highly significant. The calculated value for *r* was 0.74 (S^{**}).

Similar calculations for zinc in *Q. acutifolia* would have shown that *cv* in soils (31%) implied a small range for which the threshold value of the *cv* ratio is 1.2.

TABLE 14-2 Statistical Tests to Establish the Physiological Role of Elements in Plants and to Predict the Significance of the Plant–Soil Correlations for These Elements

| | Range of concentration of element in soil in terms of *cv* | | | |
| | Large range $cv \geq 70\%$ | | Small range $cv \leq 70\%$ | |
Role of element	Ratio of *cv* (plant/soil)	Predicted significance of *r* (plant–soil)	Ratio of *cv* (plant/soil)	Predicted significance of *r* (plant–soil)
Essential	< 0.60	$P > 0.001$	< 1.2	$P > 0.001$
Nonessential	≥ 0.60	$P \leq 0.001$	≥ 1.2	$P \leq 0.001$

r = correlation coefficient.
cv = coefficient of variation.
P = level of probability.

SOURCE: Timperley et al. (1970b).

The actual ratio was 0.68, implying that zinc was an essential element for this plant. A nonsignificant value of r for the plant–soil relationship would have been expected. The actual value was −0.08 (NS).

14-4   CONCLUSIONS

If any or all of the above tests are carried out after a biogeochemical orientation survey, it will be possible to predict with reasonable certainty the expected behavior of the species investigated. These tests are of course not infallible but would be useful as screening procedures in which the more unpromising species would be eliminated. Other cases are better evaluated by calculation of the correlation coefficients (see Chapter 13) for the plant–soil relationships of an elemental content. If computer facilities are not available for calculating values of r, the above tests will be an acceptable, though inferior, substitute.

The findings of Timperley et al. (1970b) reported in this chapter are based on seven New Zealand species. All species tested showed the above trends without exception. There is obviously no guarantee that these results will apply equally well to other species in other countries and it will probably be necessary to carry out research in this direction in other parts of the world. It is quite probable however, that the basic principles enumerated here are of universal application. It may only be necessary for other workers to find their own threshold values for "large" and "small" concentration ranges of copper, zinc, and other elements in soils.

The above findings concerning the "difficult" elements copper and zinc will not necessarily contradict the large amount of biogeochemical data accumulated in the literature concerning the successful application of biogeochemical prospecting methods to these two elements. It may merely be, as previously stated, that the failure rate for these two elements is higher than for other elements because of the difficulties enumerated above.

SELECTED REFERENCE
General: Timperley et al. (1970b)

CHAPTER FIFTEEN
CHEMICAL
ANALYSIS
OF PLANTS
AND
SOILS

During the past decade, progress in instrumental methods of analysis has greatly outstripped that of previous years and this has had far-reaching effects on those methods of mineral exploration that require chemical analysis. The development of techniques that make it possible to carry out analyses cheaply, accurately, and at great speed has had a marked stimulatory effect on geochemical and biogeochemical prospecting, and there is little doubt that the current level of activity in these fields could not be maintained without these instrumental methods.

Although it is not true to say that the days of "wet chemistry" are numbered, it certainly is true that the working time of the analyst is becoming increasingly occupied with mere preparation of samples for subsequent instrumental analysis. This development has also been beneficial to mineral exploration since there has been a corresponding reduction in the skill needed by the analyst so that a small number of semiskilled personnel can now undertake a greater volume of work than could a larger number of highly trained analysts in former days.

Since instrumental methods of analysis have become so important, the emphasis in this chapter will be on two of these: atomic-absorption spectrophotometry and emission spectrography. Other techniques, such as X-ray fluorescence and colorimetry, will also be mentioned although they are not used as extensively as the first two methods.

The analysis of material for geochemical or biogeochemical prospecting involves a combination of all or some of four basic steps: the preparation and dissolution of the sample followed by the separation and determination of the element or elements. Whereas all methods of analysis involve the first and last of these steps, most modern instrumental techniques do not require a preliminary separation of the analysis element and others do not require the dissolution stage. Figure 15-1 gives a schematic representation of the various processing stages required for three important instrumental methods of analysis and for classical "wet chemical" procedures. The term wet chemical includes general methods such as colorimetry, fluorimetry, gravimetry, and titrimetry. Figure 15-2 shows the usual working

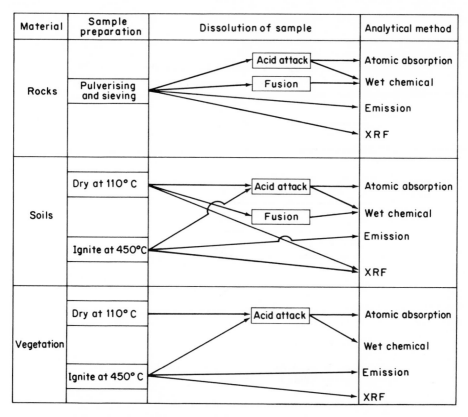

FIGURE 15-1 Schematic representation of preparation, dissolution, and analysis of rocks, soils, and vegetation.

ranges of the various analytical procedures.   Since biogeochemical methods of prospecting are usually concerned with trace elements, it is only the more sensitive of the techniques shown in Figure 15-2 which are of general application for this method of prospecting.

15-1   SAMPLE PREPARATION
          The question of sample preparation has been mentioned briefly in Chapter 13 but further discussion of this topic is justified at this stage. The type of sample preparation depends on the analytical method to be used. For emission spectrography, soils or plant material ignited at 450°C are required. For X-ray fluorescence, plant ash or oven-dried soils may be used and the ignition stage for soils (although desirable), is not absolutely essential. For most other methods of analysis, a solution will be required.

          It has frequently been asserted (e.g., Carlisle and Cleveland, 1958) that a disadvantage of biogeochemical prospecting methods compared with soil-sampling procedures, is the necessity for an additional step (ashing of the plant material)

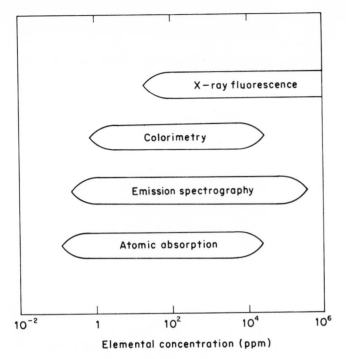

Elemental concentration (ppm)

FIGURE 15-2 Working ranges of elemental concentrations in samples analyzed by various methods.

in the analytical sequence. But inspection of Figure 15-1 will show that ignition of soil samples is also required if the commonly used technique of emission spectrography is to be employed. A further factor to be considered is that it is considerably easier to dissolve plant ash than soils or rocks so that the combined ignition-dissolution stage for plant material may need less operator time and will probably be more effective in solubilizing the constituents than will the single dissolution step for rocks or soils.

15-2   SAMPLE DISSOLUTION BY ACID ATTACK

SOILS AND ROCKS   It is more difficult to dissolve soils and rocks than it is to dissolve plant ash because many of the inorganic constituents of rocks or soils are bound to various components of the sample with the forces of very different magnitudes. Warren and Delavault (1956) have discussed this problem and have mentioned that elements may be found in soils in at least six forms: free metal ions extractable with water; ions adsorbed on clay minerals and extractable with saline solutions; metals chelated to organic material; metals in secondary oxides or carbonates which are extractable with mild acids; metals in primary silicates or oxides which need attack with strong acids; metals found as free sulfides and which require very strong attack with aqua regia or other acids. Lakin et al. (1952) have suggested several acid mixtures capable of dissolving elements bonded in all of the six ways mentioned above.

The sample dissolution stage is a very great source of error in the analysis of rocks and soils. This is demonstrated in Table 15-1 which shows values for copper in chalcopyrite and nickel in stream sediments. Depending on the nature, duration, and temperature of the acid attack, it is obviously possible to obtain a range of values spreading over two orders of magnitude. This problem is frequently aggravated by a lack of communication between the field worker and analyst. All too often, samples will arrive at the laboratory with the cryptic comment "do Cu by a.a." Unless the analyst uses exactly the same method of dissolution as previously employed for the same field worker, the results may not be comparable. In practice, provided a standard method is used in dissolving samples for the whole of an exploration survey, few problems will arise even if transfer of the metals to the aqueous phase is not complete.

In field tests where it may not be feasible to carry out hot digestion of samples with strong acids, use is made of cold extractants such as hydrochloric or nitric acids and aqua regia. As mentioned above, it is not particularly important that extraction is incomplete provided that the degree of extraction remains relatively constant for the whole of the exploration survey.

Probably the most efficient method of acid attack is a combination of hot aqua regia and hydrofluoric acid. One problem with this method of attack is that neither glass nor platinum ware will withstand the acids used. Even if aqua regia is

TABLE 15-1   Effect of Different Methods of Digestion on the Apparent Copper Content of Chalcopyrite and the Nickel Content of a Stream Sediment

Sample	Temperature and time of digestion	Acid used	Apparent elemental content (ppm)
		NICKEL	
Stream sediment	100°C	12 M perchloric	20
		4 M nitric	60
		aqua regia	150
		5 M hydrochloric	320
		COPPER	
Chalcopyrite	20°C for 2 min	2 M perchloric	300
		2 M hydrofluoric/perchloric	300
		2 M hydrochloric	1,200
		2 M nitric	6,000
	50°C for 15 min	2 M hydrofluoric/perchloric	300
		2 M perchloric	650
		2 M hydrochloric	1,300
		2 M nitric	21,000
	100°C for 15 min	2 M perchloric	1,500
		2 M hydrochloric	3,500
		2 M hydrofluoric/perchloric	40,000
		2 M nitric	55,000
		5 M nitric/hydrochloric	130,000
	180°C for 60 min	9 M perchloric	200,000

SOURCES: Bondar (1969); Sampey (1969).

replaced with hydrochloric or nitric acids, the problem of attack of the glassware remains as long as hydrofluoric acid is included. Brooks (1960) has described the use of Teflon beakers for dissolving rocks or soils with a mixture of aqua regia and hydrofluoric acid. Although these containers have the advantage that they may be placed directly on a hot plate, they are not really practicable for very large numbers of samples due to their excessive cost. The procedure used in the author's own laboratory involves the use of polypropylene beakers for the acid attack. Although these containers are not completely resistant to aqua regia and hydro-fluoric acid, they will withstand several treatments with this mixture and, since they are inexpensive (7¢ each), they can be discarded when signs of damage become evident. The complete details of acid attack are as follows.

Weigh about 0.2 g of finely ground (−80 mesh) soil or rock into a 50-ml squat polypropylene beaker and add 5 ml of a 1:1 mixture of hydrofluoric acid and aqua regia. Evaporate to dryness on a specially constructed water bath. The lower part of the bath is of conventional design but the lid comprises a sheet of $\frac{1}{8}$-inch Teflon with holes cut to fit the beakers. The holes are of such a size that the beakers can be pressed some distance into the Teflon so that only about $\frac{1}{2}$ inch protrudes above the surface of the lid. When the contents of the beaker are dry, add 10 ml of 2 M hydrochloric acid and replace in the water bath for about 10 min until the residue has dissolved in the acid. Weigh the beaker and contents before and after addition of the acid in order to obtain the dilution factor. After filtration or decantation the solution is ready for analysis.

A procedure commonly used in commercial laboratories is to take 2 g of rock or soil and take to dryness with a mixture of 5 ml of concentrated nitric acid and 10 ml of perchloric acid (70%). The procedure is repeated and the residue is dissolved in dilute perchloric acid. Another procedure involves two successive attacks with 15 ml of aqua regia. In both of these methods of acid attack, the silicates are not completely destroyed but solubilization of trace elements is relatively complete.

PLANT ASH The dissolution of plant ash is considerably easier than that of rocks and soils since most of the constituents of the ash are easily soluble in hot 2 M hydrochloric acid. A recommended procedure is as follows:

Weight about 0.1 g of plant ash into a 12 mm × 150 mm test tube and add 10 ml of 2 M hydrochloric acid. Place the test tube in a beaker of boiling water for 15 min and decant or filter the residue unless it settles readily. Solubilization of the trace constituents is usually almost 100% complete. If silver or lead are to be analyzed, the ash should instead be dissolved in 2 M nitric acid.

15-3   SAMPLE DISSOLUTION BY FUSION

Fusion of samples of rocks and soils with a fivefold amount of fusion mixture (1:1 sodium and potassium carbonates) is an effective way of achieving dissolution. The mixture is heated in a platinum crucible and extracted with hot water. The procedure is unnecessary for plant ash, but essential for rocks and soils when analyzing for such elements as tin or tungsten which are difficult to dissolve by acid attack.

15-4   ATOMIC-ABSORPTION SPECTROPHOTOMETRY

THEORY   The development of instrumental methods of atomic-absorption spectrophotometry by Walsh (1955) has been without doubt one of the greatest advances in analytical chemistry in recent years.   The technique is of great simplicity and is relatively free of serious interference problems so that analyses for many elements can be carried out in the same solution without the need for tedious separation procedures.

The basic theory of atomic absorption was known as long ago as the middle of the nineteenth century.   In essence, it concerns the observation that certain wavelengths of radiation emitted by *excited atoms* (*emission spectrum*) will be strongly absorbed by *unexcited atoms* of the same element.   Hence, radiation emitted by these excited atoms will be reduced in intensity if allowed to pass through a region in space containing a population of unexcited atoms of the same element.   A spectral line absorbed in this manner is said to be a *resonance line*.   The extent of the attentuation of the emission spectrum is determined by a detector and is a direct measure of the concentration of the analysis element.

The emission spectrum of an element is formed in the following way.   When the atoms become excited, their planetary electrons are promoted to higher energy levels and when these atoms lose their energy by collisions or other processes, the electrons return to lower energy levels.   The resultant emission of energy (E) produces radiation of a particular wavelength in accordance with the formula:

$$E = \frac{hc}{\lambda}$$ (15.1)

where h is Planck's constant, c is the velocity of light, and λ is the wavelength of the spectral line emitted.

The unexcited atoms capable of absorbing radiation are said to be in the *ground state* where the electrons occupy the lowest possible energy levels.

Figure 15-3 shows an absorption and emission spectrum for lead.   It will be noted that absorption lines are fewer than emission lines and of different intensities. Emission lines are used in emission spectrography and absorption lines in atomic-absorption spectrophotometry.   It is therefore important to know the relationship between them in order to understand the principles of the instrumentation involved. The intensities of both sorts of lines will depend on the relative populations of excited and ground-state atoms in the sample to be analyzed.   The following equation illustrates this relationship.

$$\frac{N_e}{N_g} = ke^{-E/KT}$$ (15.2)

Where k and K are constants, E is the energy involved in the electronic transition, T is the temperature of the system, and N_e and N_g are the respective populations of excited and ground-state atoms.

In equation (2) the constants are of such dimensions that, even at the very high temperature of the dc arc (ca. 5000°C), the population of ground-state atoms is far higher than that of excited atoms.   Consider as a specific example the spectral

Absorption lines

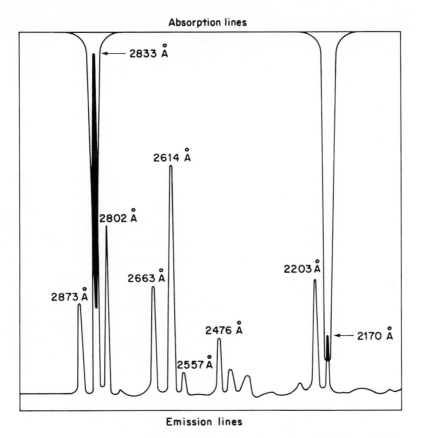

FIGURE 15-3   Absorption and emission spectral lines for lead.

line Zn 2138. There are a million times more zinc atoms in the ground state than in the excited state at 5000°C and it is therefore obvious that for this element an analytical procedure based on a population of ground-state atoms will be much more sensitive than one that relies on excited atoms. In fact, atomic absorption is about 10,000 times more sensitive than emission spectrography for the determination of zinc.

INSTRUMENTATION   Atomic-absorption spectrophotometers comprise four main components. These are: the source, absorption unit, monochromator, and detection system.

Sources comprise hollow cathode lamps which are usually designed for a single element although multielement lamps are also available. The emission spectrum is produced by passing a dc discharge which ionizes the low-pressure gas (argon or neon) in the lamp. The gas ions strike a hollow cathode manufactured from the analysis element and release a cloud of excited atoms which produce the emission spectrum.

The absorption unit comprises a nebulization system which sprays the analysis solution into a long narrow flame fed with a mixture of air-acetylene, nitrous oxide-acetylene, or other combination of gases. The purpose of the flame is to decompose the chemical compounds into atoms and provide a long absorption path. The temperature must be high enough to decompose compounds into atoms and yet not so high that a significant proportion of the ground-state atoms becomes excited.

The monochromator (usually of the grating type) serves to isolate the particular spectral line which is to be used for the analysis.

The detector system comprises a circuit to amplify the signal from a photo-multiplier which receives the energy of the spectral line from the monochromator. Alternating-current amplification systems are preferable because of their inherent stability and are a feature of most commercial instruments. The result is obtained as a digital readout, meter reading, or chart recording.

Figure 15-4 gives a schematic representation of an atomic absorption spectro-photometer. The operation of the instrument is very simple and consists of nebulizing analysis solutions into the flame and comparing the absorptions with standard solutions run for each batch of analyses.

Since commercial instruments became available about 1960, a large number of different models have been produced. In the early days of development, instruments soon became obsolescent due to the very rapid advances made. At present however, it is possible to buy an instrument with the reasonable certainty that it will not become obsolete for several years. Present developments include *resonance detectors* (specific for one element only) and systems for simultaneous analysis of several elements by using multipassage through a single flame and incorporating monochromators or resonance detectors. MacLiver et al. (1969) have described an instrument of this type and reported that 100 samples per hour could be analyzed for the 6 elements: cobalt, copper, lead, nickel, silver, and zinc.

PERFORMANCE OF ATOMIC-ABSORPTION SPECTROPHOTOMETERS Catalogs and brochures issued by manufacturers of atomic-absorption equipment are not necessarily the most objective sources of information on the performance of these instruments. Several books on the subject have, however, been written

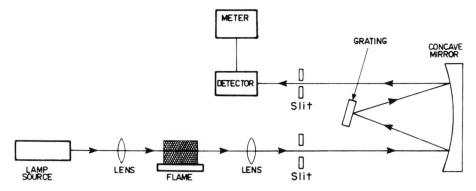

FIGURE 15-4 Schematic representation of an atomic-absorption spectrophotometer.

TABLE 15-2   Detection Limits of the Elements for Various Methods of Analysis

Element	Atomic absorption	Atomic fluorescence	Colorimetry	Emission analysis	X-ray fluorescence
Ag	0.005	—	0.05	0.05	8
Al	0.12	—	0.005	2	40
As	0.58	—	0.10	100	3
Au	0.01	—	0.05	10	16
B	6	—	0.50	10	—
Ba	0.10	—	1	5	13
Be	0.002	—	0.08	10	—
Bi	0.02	—	6	20	19
Ca	0.002	—	1	2	4
Cd	0.001	0.0001	0.03	10	9
Co	0.007	—	0.03	10	2
Cr	0.005	—	0.07	1	2
Cu	0.005	0.04	0.02	0.5	5
Fe	0.01	—	2	5	2
Ga	0.07	10	0.4	3	3
Ge	1	—	0.2	5	3
Hg	0.20	0.10	0.05	100	17
In	0.05	10	0.5	1	10
Ir	4	—	0.06	50	15
K	0.005	—	2	2	20
La	8	—	40	10	12
Li	0.005	—	0.03	0.5	—
Mg	0.0003	—	0.02	2	300
Mn	0.005	—	0.05	10	2
Mo	0.010	—	0.10	5	5
Na	0.005	—	80	0.5	600
Nb	20	—	0.05	30	4
Ni	0.01	1	0.04	5	2
Pb	0.01	—	0.06	5	18
Pd	0.50	—	0.10	10	7
Pt	0.50	—	0.10	50	15
Rh	0.03	—	0.50	10	7
Ru	0.30	—	1	10	6
Sb	0.20	—	0.04	20	11
Sc	0.20	—	0.08	2	4
Se	0.50	0.15	2	—	3
Si	0.1	—	1	20	35
Sn	0.1	—	0.6	10	11
Sr	0.01	—	0.10	5	3
Ta	6	—	0.40	100	11
Te	0.30	0.05	4	200	11
Ti	0.10	—	0.1	10	2
Tl	0.20	0.04	0.2	1	18
U	12	—	3	100	30
V	0.02	—	0.1	5	2
W	3	—	0.3	20	12
Y	0.30	—	10	10	12
Zn	0.002	0.0001	1	100	3
Zr	5	—	0.5	10	4

N.B. All units are ppm.

SOURCES: Ahrens and Taylor (1961); Koirtyohann (1967); Morrison and Skogerboe (1965).

(Robinson, 1966; Elwell and Gidley, 1966; Slavin, 1968) and are useful as more objective sources of information.   Table 15-2 lists the limits of detection of four instrumental techniques of analysis.

The question of sensitivity is of great importance for biogeochemical and geo-chemical prospecting since care must be taken that the correct method of analysis is used at all times.   For example, a field worker used to having his samples analyzed by atomic absorption might request that molybdenum be determined by this technique.   If, as would probably be the case, the results were to lie near the method's limits of detection, the accuracy of the data would be in question and the field worker would have been better advised to have asked the analyst to use a different method such as emission spectrography.   Inspection of Table 15-2 would seem to indicate that atomic absorption is in general much more sensitive than emission analysis.   This does not, however, take into account the dilution factor. The figures quoted for atomic absorption should be multiplied by a factor of 50 to make them comparable with emission spectrography which uses solid samples. This factor assumes that total solids in the solution will not exceed 2%.   It is possible to make solutions more concentrated than this—with the attendant risk of increasing interference problems.   It should also be realized that the figures quoted in Table 15-2 are optimum values obtained under carefully controlled con-ditions and that these levels will seldom be obtainable under working conditions.

The sensitivity of atomic-absorption instruments can be improved by increasing the amplification of the signal.   But this also increases the "noise" level and is self-defeating unless a very stable signal is received from the source.   Double-beam commercial instruments are available to increase the stability of the signal and are more sensitive than single-beam units because greater amplification can be used. With the increasing stability of modern hollow cathodes, however, the advantages of double-beam instruments are less than they were previously.

AN EVALUATION OF ATOMIC-ABSORPTION SPECTROPHOTOMETRY AS AN ANALYTICAL TOOL FOR BIOGEOCHEMICAL PROSPECTING   During the past decade, the growth of atomic absorption as an instrumental technique has been phenomenal.   In 1968 there were already 3000 instruments in use in the United States alone and these represented a capital outlay of over $20,000,000 (Lewis, 1968).   This number of instruments is expected to double by 1971.   When instruments first become commercially available, the result of initial overenthusi-asm was an overrating of the capabilities of the technique; but now, after a decade of experience, it is possible to place atomic absorption in proper perspective.

The method certainly possesses the advantages of specificity, sensitivity, relative freedom from interferences, multielement determinations on the same sample, multielement determinations with the same instrument, freedom from elapsed time requirements, and the provision of data in direct readout form.   There are, however, some disadvantages, which have only been appreciated fairly recently.   Ward and Nakagawa (1969) have shown that quite severe matrix effects can be produced in solutions of high ionic strength.   To some extent this problem can be overcome by dilution of the sample, but this in turn reduces sensitivity.   Even in dilute solutions, other interference effects are present and the above authors have suggested that the ultimate solution to the problem is to have standards with the same matrix as the analysis solutions.

Despite the cautionary note of the above discussion, there is no doubt that atomic absorption is proving of immense value for geochemical and biogeochemical exploration and is at present probably the most favored method of analysis used for these procedures.

15-5   EMISSION SPECTROGRAPHY
Emission spectrography is one of the oldest of the instrumental methods. It dates back to 1861 and the spectroscopic discovery of cesium by Bunsen and Kirchhoff. The technique remained qualitative or at best semiquantitative until the pioneering work of Goldschmidt and his co-workers at Göttingen in the early 1930s. After this period, there was a rapid spread in use of the technique and it became widely used for the analysis of a large number of elements in rocks, soils, and plant material. Until recently, emission spectrography was the most favored instrumental method of analysis used for prospecting purposes and with colorimetry accounted for nearly all analyses made.

THEORY   The theory of emission spectrography basically resembles that of atomic absorption. Emission lines are produced according to Eq. (15.1). But, as can be seen from Eq. (15.2), even under favorable conditions there are far fewer excited atoms than ground-state atoms. Thus, it is not surprising that emission analysis is inherently less sensitive than atomic absorption even if allowance is made for the dilution effect referred to above. The theory and practice of emission spectrography have been described by Ahrens and Taylor (1961).

INSTRUMENTATION   Instrumentation for emission spectrography consists basically of three components: the excitation source, the optical system, and the detection system.

Excitation is usually carried out in the dc arc, although ac arc or spark sources are also often used. Samples are loaded into carbon or graphite electrodes. The image of the arc is focused at the slit of the spectrograph or at the collimator. The optical system of the spectrograph includes a quartz (or glass) prism or diffraction grating in order to split the arc image into its spectral lines.

The detection system consists of photographic plates for optical spectrographs or electronic equipment for direct-reading instruments. The individual spectra appear in photographic plates as a series of dark lines whose pattern characterizes the element and whose blackness, measured on a microdensitometer, is a function of the concentration of the element.

A schematic representation of the optical arrangements of prism and grating spectrographs is shown in Figure 15-5. There is some advantage in using a grating spectrograph, as dispersion remains constant through the whole spectral range whereas in prism spectrographs, dispersion decreases with increasing wavelengths.

In quantitative analysis it is customary to add to the sample a definite proportion of an *internal standard*, such as palladium, which is rare enough not to be

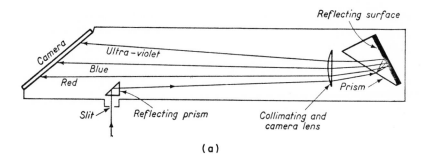

(a)

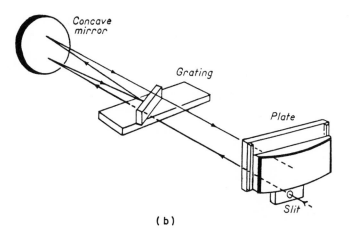

(b)

FIGURE 15-5 Diagrams of optical systems for emission spectrography. (a) Prism-type instrument. (b) Grating instrument. From Ahrens and Taylor (1960). By courtesy of Interscience Publishers.

found in significant amounts in the sample and which at the same time volatilizes in a similar manner to the analysis elements. The ratio of the blackness of the analysis line to that of the internal standard is compared with standards containing the same proportion of internal standard and the concentration of the element is calculated from this ratio.

The measurement of the blackness of spectral lines is very tedious. An acceptable short cut for semiquantitative work is to rotate a *step sector* in front of the slit of the spectrograph during the arcing. This has the effect of splitting the spectral line into a graded series of steps (usually seven) each representing a radiation intensity half that of the preceding step. For a seven-step sector the first step will obviously represent a blackness 64 times greater than that of the seventh. In biogeochemical prospecting work, provided that there is good contrast between background and mineralized values, it is sometimes merely sufficient to designate the highest step number that is just visible and take this as being a very approximate value for the concentration of the element.

Direct-reading spectrographs are a much superior and speedier alternative to the optical spectrograph but are also much more expensive. The source and optical arrangements are similar to those of optical instruments but the detection system incorporates a number of photomultipliers positioned at various exit slits and corresponding to specific analysis lines. Electrical impulses from the photomultipliers are stored in condensers during excitation and are released sequentially at the end of the exposure. The result is displayed on a scale, printed on a typewriter, or punched on data cards for computer processing. Spectrographs are now available with provision for up to 22 channels so that this number of elements can be determined simultaneously.

PERFORMANCE OF EMISSION SPECTROGRAPHS   Optical spectrographs cannot compete either in speed, precision, or costs with atomic-absorption spectrophotometers when a limited number of elements are to be analyzed. Their use is, however, justified for a number of elements such as boron, beryllium, and molybdenum which are more easily determined by emission spectrography than by other instrumental techniques. One great disadvantage of emission analysis is that line intensities of most elements vary greatly with the bulk composition of the sample. This *matrix effect* can be reduced by having standards of the same bulk composition as the analysis samples. Table 15-2 includes data on the limits of detection of elements determined by emission spectrography. A further disadvantage of optical instruments is that they demand a higher level of skill from the analyst than that required for direct readers or atomic absorption.

Direct-reading instruments are a useful alternative to atomic-absorption units as an analytical tool. The high capital cost of these instruments precludes their use for small numbers of samples, particularly if a limited number of elements is to be analyzed.

The reproducibility of direct-reading instruments is comparable with that of atomic-absorption equipment ($\pm 1\%$). For optical instruments, a coefficient of variation of ± 5–10% might be expected for replicate analyses of identical samples, which is, however, adequate for biogeochemical prospecting work.

15-6   X-RAY FLUORESCENCE SPECTROGRAPHY

X-ray fluorescence spectrography is a nondestructive method of analysis but has not been used very widely for the analysis of plant material, rocks, or soils because of certain basic factors. These are the high capital cost of equipment, the relative slowness and insensitivity of the method, and the fact that relatively large samples are required for analysis. This latter factor is particularly disadvantageous for the analysis of plant ash when samples tend to be small. X-ray fluorescence does have the advantage that simultaneous determinations of many elements can be carried out. Relatively precise values can be obtained for the concentrations of the major constituents of rocks, soils, or plant ash. Table 15-2 lists the limits of detection of the elements by this method and demonstrates its relative insensitivity.

The basis of X-ray fluorescence is the bombardment of the sample with *primary X-rays* produced from a tungsten or molybdenum X-ray tube. These primary X-rays expel electrons from the atoms in the sample leaving a "hole." This vacancy is filled by other electrons from higher energy levels and the net result is a characteristic X-ray emission spectrum according to the principles of Eq. (15.1). The *secondary X-rays* thus emitted pass through a rotating crystal where, by Bragg's law, only certain wavelengths are diffracted for a specific angle between the X-ray beam and the axis of the crystal. The diffracted X-rays are recorded by a detector (Geiger counter or scintillometer) rotating at twice the rate of the crystal and the output of the detector is plotted automatically as a function of the angle of rotation. From the magnitude of this angle, the nature of the element can be determined; its concentration is calculated from the strength of the X-rays received by the detector.

As mentioned previously, X-ray fluorescence is hardly applicable for speedy analysis of large numbers of samples. It can be useful in providing a total analysis of the major constituents of the material in a fraction of the time that would be required by "wet chemical" methods.

15-7   OTHER METHODS OF ANALYSIS

COLORIMETRY   With the advent of atomic absorption and direct-reading spectrographs, there has been a rapid decline in colorimetric methods of analysis applied to mineral exploration. Nevertheless, the pioneering work of Sandell (1959) and his co-workers is still useful for the colorimetric analysis of elements such as tin, tungsten, etc., which are difficult to analyze by instrumental methods when the concentrations are low. There will also be a continuing need for colorimetric methods in field analyses although, even here, the demand will diminish with a trend to miniaturization of instruments such as X-ray fluorimeters. Light-weight gamma-source X-ray units are now available at moderate cost (Watt, 1967), and it may be expected that some years from now they will be refined to the point that a considerable proportion of field analyses for the more abundant elements will be carried out by their use.

ATOMIC FLUORESCENCE   Atomic fluorescence is a new instrumental technique developed by Winefordner and Vickers (1964). The basic equipment needed is similar to that used for atomic absorption except that the source is a high-intensity electrodeless discharge tube held at right angles to a conical flame cell. When a solution containing the analysis element is nebulized into the flame, the atoms fluoresce (*resonance fluorescence*) and emit radiation of the same wavelength as was absorbed from the source. The signal passes through the monochromator and is treated exactly as for atomic absorption.

As atomic fluorescence is a type of flame emission analysis, it is possible to increase the sensitivity considerably by amplification of the signal or by increasing the intensity of the source. Limits of detection of some elements by use of this technique are also shown in Table 15-2.

Atomic-fluorescence attachments for atomic-absorption equipment are now becoming commercially available and substantial progress can be expected in the future. One of the advantages of the technique is that the working range of elemental concentrations is much greater than for atomic absorption. The sensitivity of the method for some elements, such as cadmium, mercury, selenium, tellurium, thallium, and zinc, is much greater than for atomic absorption although for most other elements there is little or no advantage. It does not seem that the technique will ever replace atomic absorption in prospecting work although it may be a useful adjunct for some elements.

RADIOMETRIC METHODS   Radiometric methods of analysis involve the measurement of alpha, beta, or gamma radiation from the materials studied. Great care must be taken in the evaluation of data from radiometric measurements since so many isotopes are capable of producing radioactivity. The usual purpose of radiometric measurements in mineral exploration is to detect uranium, but unfortunately this element has a relatively low radioactivity due to its long half life ($4.5 \times 10^9 y$ for uranium-238). Most of the radioactivity in plants or soils is derived from uranium daughters such as Ra-226 or from other radioactive series. Provided that samples are to some degree in radioactive equilibrium, measurement of daughters is a good indication of the presence of uranium.

The measurement of alpha activities in plants and soils has been used in exploration work by Anderson and Kurtz (1956), Goldsztein (1957), and Whitehead and Brooks (1969a, b). Whitehead and Brooks (1969a) compared the method with other techniques for detecting uranium and concluded that the alpha activity of plant ash is a reliable indication of mineralization.

The beta activity of soils and plants is far less reliable as an indication of uranium mineralization due to the presence of large amounts of potassium-40 in plant ash and, to a much lesser extent, in soils. Figure 15-6 shows the alpha activity of plant ash and soils expressed as a function of the beta activity, which decreases to a limiting value of about 300–400 counts/100 min/25 mg as alpha activities decrease to zero (Whitehead and Brooks, 1969a). Grodzinsky and Golubkova (1964), in making the same observations, concluded that 45–71% of the beta activity of plant ash is due to radioactive potassium.

If measurements are made by gamma spectrometry, the gamma activity of plants and soils is an extremely reliable indication of the type of mineralization present. This is an instrumental technique whereby individual contributions to radioactivity can be determined and the sources characterized. Whitehead and Brooks (1971b) have discussed the origins of radioactivity in some New Zealand plants and soils and have shown that gamma spectrometry can be used to characterize the isotopic sources of all three types of radiation. Unfortunately, the technique is at present extremely expensive and not readily available. No doubt, if a sufficient demand for these instruments develops, cheaper units will be manufactured.

Another radiometric technique used in biogeochemical prospecting is *neutron activation*. This technique is of very great sensitivity but is slow and not really suited to the analysis of large numbers of samples. The method consists of irradiating samples in a nuclear reactor. Absorption of neutrons converts the constituents of the plant ash or soil to radioisotopes whose activity can be measured by gamma spectrometry. Lobanov et al. (1966, 1967, 1967) have used the method for the determination of gold in plants, rocks, and soils.

FLUORIMETRY   There are a few elements which can only be determined satisfactorily at low levels by one or two methods. A good example of this is

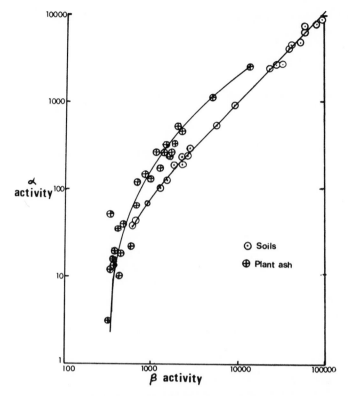

FIGURE 15-6 Alpha activity of soils and plants as a function of their beta activity (units expressed in counts per
100 min. per 25 mg sample).   From Whitehead and
Brooks (1969a). By courtesy of the Economic Geology
Publishing Co.

uranium, which is determined by measuring the ultraviolet fluorescence of a bead
formed by fusion of the sample with sodium fluoride (Grimaldi et al., 1952).
Brooks and Whitehead (1968) have described a simple fluorimetric attachment for
an atomic-absorption spectrophotometer and have used the instrument for the
analysis of uranium in plants and soils.   Unfortunately, the fluorimetric method
is very sensitive to quenching effects and the uranium should be purified by simple
solvent extraction procedures (Grimaldi et al., 1952; Nemodruk and Voronitskaya,
1962); otherwise, results will be erratic.

SPECIFIC METHODS FOR ANALYSIS OF VOLATILE SUBSTANCES   Two
methods specifically designed for volatile substances will now be discussed briefly.
One of these involves the analysis of gases from plants and soils in the course of
exploration for petroleum. The technique of *gas chromatography* is used for this
purpose (Smith and Ellis, 1963) and enables various constituents of petroleum in
plants and soils to be analyzed.
    The analysis of mercury in plants, rocks, or soils can be used as a pathfinding

technique for sulfide minerals in which mercury commonly occurs. Mercury is so volatile that it can provide a dispersion halo many times greater than the sulfide body itself. Because of its volatility, this element presents an analytical problem, which can be solved by use of a flameless atomic-absorption technique described by Vaughn (1967). In this method, the sample is heated in a quartz tube at 500°C and the mixture of mercury and volatile organic compounds is passed over silver wire which amalgamates with the mercury and allows the organic gases to escape out of the system. The mercury vapor is then thermally released into an absorption chamber where the ground-state atoms absorb the mercury line at 2540 Å from a mercury lamp acting as a source. The absorption is recorded by a detector.

15-8   CHEMICAL FIELD TESTS

Field methods used for the analysis of plants, rocks, and soils are nearly all colorimetric and are essentially semiquantitative. Many of the procedures were developed by the U.S. Geological Survey (Lakin et al., 1952) and by the Applied Geochemistry Department at Imperial College (Stanton, 1966). The accuracy, sensitivity, and field performance of these methods have been reviewed by McCarthy (1959) and Tooms (1959). Methods specifically designed for plant ash have been described by Reichen (1951), Reichen and Lakin (1949), and Reichen and Ward (1951). Most techniques designed for soils can be used for plant ash with minor modifications. One method of doing this is to digest the plant sample with a 1:5 mixture of perchloric and nitric acids heated over an alcohol burner or gasoline stove. An alternative is to ash the vegetation first and then dissolve the residue in acid. Table 15-3 lists the sensitivity of some well-known colorimetric field tests for various elements and shows the normal range of elemental background levels for soils and plant ash. In some cases the sensitivity of these methods is not sufficient to determine the lowest concentrations of elements in background samples. This is not a particularly serious problem as anomalous values are not likely to be below detection limits.

TABLE 15-3   Sensitivities of Colorimetric Field Tests

Element	Method	Sensitivity (ppm)	Accuracy (±%)	Normal range of element In background soils (ppm)	In background plant ash (ppm)
Antimony	Rhodamine B	0.5	8–20	1	1
Arsenic	Confined Spot Test	5	15–35	1–50	1
Cobalt	2-nitroso-1-naphthol	1	15–30	1–40	1–20
Copper	Dithizone	10	15–30	2–100	70–200
Lead	Dithizone	10	8–25	2–200	50–100
Molybdenum	Dithiol	1	10–40	0.2–5	1–20
Nickel	Dimethyglyoxime	50	—	5–500	40–70
Tin	Gallein	10	—	5–25	1
Tungsten	Dithiol	4	27–34	0.5–2	1
Zinc	Dithizone	50	10–25	10–300	600–1400

SOURCES: McCarthy (1959); Tooms (1959).

The reliability of field tests has been discussed by Tooms (1959) and McCarthy (1959), who applied statistical analysis to a large number of samples collected during prospecting work. These findings are shown in Table 15-3. The reliability of a field method is a function not only of the concentration of the element in the sample but also of the analyst himself.

It is clear from the above discussion that field methods of analysis are at best only semiquantitative and are justified primarily on the grounds of on-site determinations. If, however, the contrast between background and mineralized values is small, it is quite possible that anomalies could be missed in an exploration survey relying on field tests alone.

15-9   MOBILE LABORATORIES

It is obvious that there is a great difference between the quality of data obtained from a properly equipped laboratory and from semiquantitative field tests. The advantage of greater accuracy in the first case is counterbalanced by the greater elapse of time between sampling and analysis. The solution to this problem is obviously to take the laboratory into the field.

Pioneering work in the development of mobile laboratories has been carried out in the Soviet Union (Ratsbaum, 1939) and in North America (Canney et al., 1957; Holman and Durham, 1966). Malyuga (1964) describes the use of mobile optical spectrographs with the arc powered by local supplies or by mobile generators.

In Australia and North America there has also been considerable interest in the development of mobile laboratories for prospecting work. These laboratories are usually equipped with optical spectrographs and atomic-absorption spectrophotometers. Debnam (1969) has reviewed the development of mobile laboratories and has described the advantages that would accrue if direct-reading emission spectrographs could be installed in such laboratories. The main disadvantage is the cost and size of the instruments. Progress in producing smaller and more inexpensive units with little loss in performance will ensure that they will soon be installed in mobile units.

An extremely important event in recent years has been the development of a mobile spectrographic laboratory specifically designed for biogeochemical work in Canada (Fortescue and Hornbrook, 1967). The unit consists of two 28-ft house trailers; one for sample processing and the other to serve as the chemical laboratory. The units are railed to near the operation site on flat-top railway wagons and then towed the rest of the way by a lorry or tractor. The trailers are stored in a heated garage during the winter.

In the biogeochemical laboratory a three-man crew is responsible for the collection, subsampling, and preparation of the samples while a second three-man crew is responsible for the analysis. Daily output of the laboratory is about 13 elements in each of 30 samples. A flow sheet of operations carried out by the unit is shown in Figure 15-7.

The development of a mobile laboratory designed for biogeochemical work is a praiseworthy achievement and might well be emulated elsewhere. With the potentialities afforded by atomic absorption and direct-reading spectrographs, even more efficient laboratories of this type will surely make their appearance.

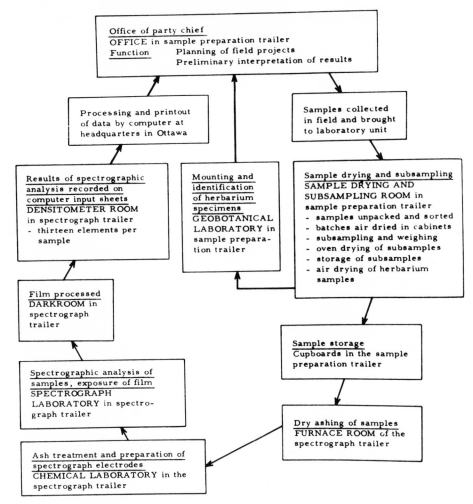

FIGURE 15-7 Flow chart for operations carried out in a Canadian mobile biogeo-chemical laboratory. From Fortescue and Hornbrook (1967). By courtesy of the Geological Survey of Canada.

15-10 AN ASSESSMENT OF METHODS OF ANALYSIS APPLIED TO BIOGEOCHEMISTRY

The success of a biogeochemical program, as well as any other project in mineral exploration, depends to a great extent on the reliability of the data and speed at which they can be obtained. Economic factors are also a consideration in choosing a suitable analytical method. The problem facing the field worker or analyst is to consider these three variables and select a technique which is the best compromise between speed, reliability, and economy.

There is little doubt that for large numbers of samples, atomic-absorption spectrophotometry and direct-reading spectrographs will have wide appeal. The

TABLE 15-4 Recommended Analytical Procedures for Various Concentrations of the Elements

Element	Concentration range of element (ppm)			
	0–10	10–100	100–1000	>1000
Ag	A	A	X	X
Al	E	E	X	X
As	C	C	A	A
Au	E	A	X	X
B	E	E	E	E
Ba	E	A	X	X
Be	E	E	E	E
Bi	E	A	A	X
Ca	A	A	X	X
Cd	A	A	X	X
Co	E	A	X	X
Cr	E	A	A	X
Cu	A	A	X	X
Fe	A	A	A	X
Ga	E	A	A	X
Ge	E	E	A	X
Hg	FA	AF	A	X
In	E	A	A	X
Ir	E	E	E	X
K	F	F	F	X
La	E	E	E	X
Li	F	F	F	F
Mg	A	A	A	X
Mn	A	A	A	X
Mo	E	E	A	X
Na	F	F	A	A
Nb	C	E	E	X
Ni	A	A	A	X
Pb	A	A	A	X
Pd	E	E	A	X
Pt	E	E	A	X
Rh	E	E	E	X
Ru	E	E	E	X
Sb	E	A	A	X
Sc	E	E	E	X
Se	F	AF	AF	AF
Si	E	E	X	X
Sn	C	C	E	A
Sr	A	A	A	X
Ta	—	C	E	E
Te	AF	AF	A	X
Ti	E	E	X	X
Tl	E	A	A	X
U	FL	FL	FL	C
V	C	E	E	X
W	C	C	E	E
Y	E	E	E	X
Zn	A	A	E	E
Zr	E	E	E	X

A = Atomic absorption.
AF = Atomic fluorescence.
C = Colorimetry.
E = Emission analysis.
F = Flame photometry.
FA = Flameless atomic absorption.
FL = Fluorimetry.

capital cost of direct-reading equipment is very high but so is its output. The cost of an optical spectrograph is less, but is output is proportionately even lower. Atomic absorption equipment requires a relatively modest capital outlay and has a high rate of output for individual elements. Colorimetric methods cannot compete with any of the instrumental techniques since they are so slow (an output of 80 analyses for one element per man day is usual). As a general rule, if large numbers of samples are to be analyzed for several elements, use of direct-reading spectrographs would be justified. For a smaller range of elements, atomic absorption would be the best procedure to use. The use of colorimetric methods is still justified where suitable instrumental methods do not exist or where the number of samples is so small that the major capital outlay for sophisticated instruments is not feasible.

Apart from the various criteria listed above, another important factor in the selection of a suitable analytical method is its sensitivity and the range of concentration values of analysis elements in the sample. Ideally, the normal working range of the analytical method should be matched with the concentration range of the element in the sample. A lack of communication between the field worker and the analyst of results in unsuitable techniques being used for analysis, with disastrous results for all concerned. For example, the cryptic message "do Mo by a.a." given by a field worker to an analyst for samples containing less than 25 ppm of this element would be a most unwise instruction since all the results would be near or below the limits of detection of the method and at best would be unreliable. The correct choice in this case would have been colorimetry or emission spectrography. At the other extreme, an instruction to "do Cu by a.a." for ore samples containing several percent of copper would be equally foolish since extensive dilution with attendant errors would be necessary in this case. Table 15-4 lists some recommended analytical procedures for various concentration ranges of the elements. The recommendations are only tentative and do not take into account such factors as volume of work and costs.

It is hoped that this review of chemical methods of analysis applied to biogeochemistry will have been instrumental to some small degree in bridging the gap which still separates the field worker from the analyst. A meaningful dialogue between the two can only be of benefit to the field of mineral exploration.

SELECTED REFERENCES
Atomic-absorption spectrophotometry: Elwell and Gidley (1966)
Emission spectrography: Ahrens and Taylor (1961)
X-ray fluorescence: Liebhafsky et al. (1960)
Field tests; Stanton (1966)
Mobile biogeochemical laboratories: Fortescue and Hornbrook (1967)

CHAPTER SIXTEEN
AQUATIC
BRYOPHYTES AS A
GUIDE TO
MINERALIZATION

Although there is an extensive literature on the subject of accumulation of trace elements by bryophytes (Lounamaa, 1956; Grodzinsky, 1959; Samoilova, 1961; Lounamaa, 1965; Shacklette, 1965a,b, 1967; Whitehead and Brooks, 1969b), it is obvious that it will never be possible to use these plants for biogeochemical methods of prospecting because of two great disadvantages. The first of these is the problem of recognition of individual species of bryophytes; the second is associated with the fact that the *rhizoids* of bryophytes are always superficial and can never give evidence of mineralization at depth.

Although most of the existing information on elemental accumulation by bryophytes applies to land forms only, there is some evidence (Puustjärvi, 1956, 1958) that aquatic bryophytes are able to accumulate elements from their environment to an even more spectacular degree. It is not unreasonable to suppose, therefore, that sampling and analysis of aquatic bryophytes might give some indication of the elemental content of the environmental water and hence indirectly afford some indication of mineralization in the region from which the water is derived.

Sampling of stream waters is a well-tested method of geochemical prospecting and can be used with some success; although the technique is not as reliable as the sampling of stream sediments. Wodzicki (1959a) has tested stream-water sampling for uranium in New Zealand and has reported two major disadvantages of the technique. The first of these concerns the fact that uranium concentrations in the stream waters tested were almost at the limit of the method's sensitivity (Ostle, 1954). It was also noted that due to violent fluctuations in the water levels in the streams of the area (annual rainfall 100 in.), uranium levels in the waters were strongly dependent on flow rates.

Aquatic bryophytes behave essentially as simple ion exchangers (Puustjärvi, 1956, 1958) and moreover are very long lived (ca. twenty years or more). Over a long period, it might be expected that trace elements would be absorbed from the streams and that their concentrations in the plant material would be some measure of the "average" content of these elements in stream waters. In this way the plants would serve as indirect indicators of mineralization in the catchment areas of the streams.

The possibility of using aquatic bryophytes as a guide to mineralization has been investigated by Whitehead and Brooks (1969c) in a uraniferous region of New Zealand. A positive response to mineralization was found and the essential details of the method are presented in Section 16-2 below in the hope that it might serve as a guide to others wishing to try the method elsewhere.

16-1   SAMPLING AND ANALYSIS OF BRYOPHYTES

Aquatic bryophytes are often found in swiftly running streams draining deeply forested regions and are usually attached to rocks and stones in the stream beds. The rhizoids have little penetrating power and serve as anchors rather than means of nourishment. Clumps of bryophytes do, however, serve as natural traps for finely divided sediment in the stream waters and a major problem arises in separating the plant material from this sediment.

When samples are collected from streams, they should be washed as carefully as possible *in situ* before the material is placed in plastic bags for onward transmission. The plant material should be washed further in the laboratory and hand picked where necessary to remove portions containing the greatest amount of entrapped sediment. The cleaned material is then ashed in Pyrex beakers at 500°C for several hours and the residue is pulverized gently with a glass rod. Grading the material through a 20-mesh nylon sieve removes a large part of the residual inorganic contamination. The plant ash (-20 fraction) is then weighed.

The main difficulty in the analysis of bryophytes is to decide to what extent contamination problems are present. One way to circumvent this difficulty is to analyze the substrate or contaminants entrapped in the clumps and make an allowance for this in the final calculations. It may be assumed that the ash of bryophytes should amount to about 10% of the dry weight of the dry material and that amounts greater than this represent contamination. During the analysis of 44 bryophytes, Whitehead and Brooks (1969c) found that the mean ash weight was 16%, with 37 specimens giving values in the range 7% to 27%. The mean contamination was probably therefore 6%, and the range zero to 17%. The same authors noted that concentrations of uranium and several other elements in the substrate and trapped sediment were significantly lower than in the bryophyte material itself. The effects of contamination were therefore negligible.

Bryophyte material is usually analyzed by atomic absorption or emission spectrography. Since the silica content of bryophytes tends to be much higher than that of vascular plants, a treatment with hydrofluoric acid is often desirable if solutions are required for analysis. The dissolution procedure is as for soils (see Chapter 15).

16-2   A CASE HISTORY FROM NEW ZEALAND

INTRODUCTION   Investigations were carried out in an area of uranium mineralization located in the Lower Buller Gorge region of New Zealand on the west coast of the South Island some 15 miles east of the town of Westport.

Mineralization is found in a formation known as the Hawks Crag Breccia (Beck et al., 1958). Primary minerals are coffinite and uraninite; secondary minerals such as autunite, gummite, and torbernite occur wherever weathering has occurred.

The annual rainfall is 100 inches and the region is extremely rough and dissected with numerous steep, swiftly running streams which, apart from ridges, form the only easy access to the region through the dense indigenous forest. A map of the area is shown in Figure 16-1 and shows the streams from which bryophytes were collected.

COLLECTION AND ANALYSIS OF THE BRYOPHYTES Collection of the bryophytes was a great problem because they grew in closely intertwined clumps containing several species together. At first an attempt was made to collect the species separately, but the separation and identification of the species was so laborious and time consuming that such a procedure was completely impracticable. Ultimately, the species composition of bryophyte material was ignored completely and all samples were taken at random from various parts of the stream and treated as composites. The commonest species sampled were: *Bryum blandum, Dicranella vaginata, Distichophyllum pulchellum, Fissidens rigidulus, Lophocolea planiuscula, Plagiochila deltoides, Pterogophyllum dentatum, Riccardia* sp., and *Thamnium pandum.*

The bryophyte samples were analyzed for beryllium, copper, lead, and uranium. The first three elements were included because previous work (Cohen et al., 1969) had shown they are good pathfinders for uranium in soils and stream sediments of

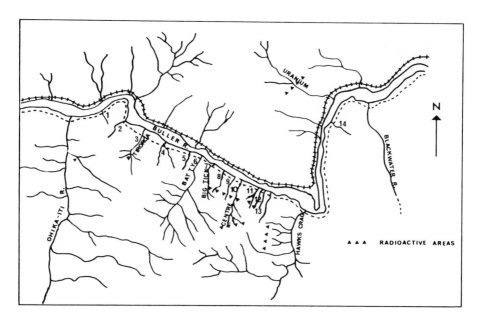

FIGURE 16-1 Map of the Buller Gorge uraniferous area of New Zealand showing streams from which aquatic bryophytes were sampled. From Whitehead and Brooks (1969c). By courtesy of the American Bryological and Lichenological Society.

the area. Beryllium was analyzed by emission spectrography, copper and lead by atomic absorption, and uranium by fluorimetry. The data are shown in Table 16-1.

DETERMINATION OF URANIUM IN STREAM WATERS   The aim of the work was to compare the uranium content of bryophytes with the uranium content of the stream waters and hence indirectly with the presence or absence of mineralization in the respective catchment areas. The latter comparison was rendered possible by the fact that the area had already been investigated thoroughly geochemically (Wodzicki, 1959a,b; Cohen et al., 1967; Cohen et al., 1969), mineralogically (Beck et al., 1958), and also radiometrically by airborne gamma scintillometry.

Data for the uranium content of five of the streams sampled were available from previous work by Wodzicki (1959a) but an attempt was made to obtain further data by determining the uranium content by means of peat *adsorbers* treated in the manner recommended by Horvath (1960). Samples of New Zealand Hauraki peat (10 g) were washed successively in tap water, distilled water, ethyl alcohol, and benzene. The samples were then dried, placed in finely woven cotton bags, and anchored in each stream for a period of one week. At the end of this period, the contents of the bags were removed, dried at 110° C, ashed at 450° C, and analyzed fluorimetrically for uranium. The results of these analyses, together with data by Wodzicki (1959a), are shown in Table 16-1.

The uranium content of the peat ash could not be used to determine quantitatively the absolute amounts of this metal in the waters but was a useful indication of the relative concentrations in the stream waters.

TABLE 16-1   Elemental Concentrations in Bryophytes and Stream Waters from the Lower Buller Gorge Region of New Zealand

Stream no and name	Uranium content of peat adsorbers in ash (ppm)	Uranium content of water (ppb)	Elemental content of bryophytes (ppm in ash)				Known mineralization in watershed of stream
			Be	Cu	Pb	U	
1	10.5	—	82	37	910	14.3	None
2	18.6	3.25	69	45	296	19.5	None
3 (Tiroroa Ck.)	18.6	2.30	41	14	120	16.8	Weak
4	9.3	—	33	20	230	8.8	None
5	9.4	—	36	17	345	12.1	None
6 (Batty Ck.)	9.0	1.40	30	6	140	4.7	None
7 (Big Tick Ck.)	57.0	—	109	78	510	11.2	Weak
8 (Jones Ck.)	324.0[a]	—	33	143	260	52.0	Strong
9 (Hornfels Ck.)	15.6	—	92	42	410	12.8	Strong
10 (Centre Ck.)	18.3	—	93	30	293	18.7	Strong
11 (Robyn Ck.)	33.0	—	53	14	690	68.0	Strong
12	36.0	—	65	60	650	86.0	Strong
13	75.0	0.80	43	140	180	6.0	Weak
14	4.0	0.65	13	61	150	0.7	None
Reliability index (max. = +25)	+17	−2	+8	+3	+5	+13	—

[a] High value due to contamination of watershed by extensive mining operations.

SOURCE: Whitehead and Brooks (1969c).

In order to check the extent of contamination by entrapped material, the sediments were analyzed and found to average 3.1 ppm for beryllium, 63 ppm for copper, 54 ppm for lead, and 5.8 ppm for uranium. Since all these values (except for copper) are lower than the concentrations in bryophyte material, and since entrapped sedimentary material averaged 6% and did not exceed 17%, contamination did not appear to be a serious problem in this work.

Flow rates of the streams were measured in each case. These values were obtained by determining the cross-sectional area of the stream and measuring the velocity of a floating object such as a small piece of paper. The cross section was measured by taking the depth at various points between the banks. Flow rates are usually recorded in cubic feet per second (cusecs) and are obtained by multiplying the cross-sectional area by the velocity of the stream water measured in feet per second.

STATISTICAL INTERPRETATION OF THE DATA   Statistical interpretation of the data was rendered difficult by the relatively small number of samples involved. Of the 14 streams investigated, 8 were known to have mineralization in their catchment areas; the mineralization in two cases being weak.

A simple statistical approach to such a problem is the application of the answer "yes" or "no" to the following question: *"Did the bryophyte data correctly indicate the presence or absence of uranium mineralization in the stream catchment area?"* The well-known *Games Theory* (Neumann and Morgenstern, 1953) was applied to this problem in the following manner. All uranium concentrations in bryophytes were arranged in decreasing order of magnitude and it was assumed that the top 8 values represented mineralization (since 8 of 14 streams were believed to drain mineralized areas). It should also be noted that the eighth value in the sequence was taken as representing the threshold level of mineralization.

Each stream was now considered in turn; and if the concentration value did, in fact, confirm the known existence of strong mineralization in the catchment area, a value of +2 was assigned to the stream. If a low figure confirmed the absence of mineralization, a similar value of +2 was assigned. If, however, the uranium content did not agree with the presence or absence of mineralization, a value of −2 was assigned. For cases of weak mineralization, values of +1 or −1 were given. The maximum and minimum scores were +25 and −21, respectively.

From Table 16-1, it can be seen that a value of +13 was obtained for the uranium content of bryophytes as an indication of mineralization. The uranium content of the peat adsorbers had a high reliability index of +17. The pathfinders were far less effective in indicating mineralization and gave values ranging downward from +8 for beryllium. Although the uranium content of the 5 streams studied by Wodzicki (1959a) was apparently no guide at all to mineralization (as the author himself discovered), the number of samples was too small to be treated in a reliable manner by statistics.

Figure 16-2 shows the uranium content of bryophytes as a function of the concentration of this element in the peat adsorbers. With two exceptions, there is close agreement between the two variables expressed on a logarithmic basis.

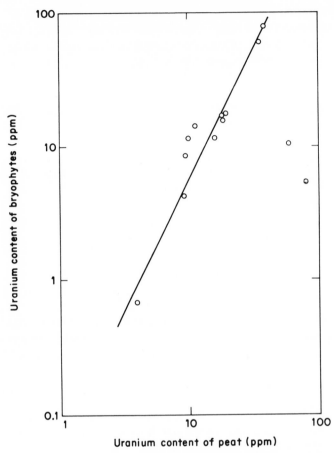

FIGURE 16-2 Uranium content of bryophytes as a function of the amount in peat adsorbers. After Whitehead and Brooks (1969c). By courtesy of the American Bryological and Lichenological Society.

16-3   USE OF AQUATIC BRYOPHYTES FOR EXPLORATION SURVEYS

In the previous section it was shown that an orientation survey apparently indicated that aquatic bryophytes could be used to detect mineralization in the catchment area of the stream concerned. The data would have been even more significant statistically if other variables could have been controlled. To be more explicit, the uranium content of bryophytes was an indication of the uranium content of the water rather than of mineralization, since a large catchment area with a relatively high uranium background concentration in the country rock could well have given a larger water anomaly than an area containing a small but highly mineralized ore body.

The flow rate of the water will also be a factor in determining the uranium content of the water. Thus, if all uranium values for bryophytes are "normalized" to a constant standard by dividing by the flow rate of the stream, a better basis of comparison can be established.

Once the existence of mineralization in the catchment area of a stream has been suspected, the tributaries should each be tested by the bryophyte method until the source of the mineralization has been discovered.

Although the above case history concerned uranium, there is no reason why the bryophyte method should not be applied to other elements, since the principles enumerated should be of wide application. The reliability of the procedure will never be as good as the analysis of stream sediments except, perhaps, in the case of uranium, which is very soluble and does not remain long in sediments. The use of bryophytes may well be an alternative or adjunct to water sampling; but since its use has been confined to the single example described above, further work will need to be undertaken before its full significance for mineral exploration can be evaluated.

SELECTED REFERENCES

Aquatic bryophytes as a guide to mineralization: Whitehead and Brooks (1969c)

Cation-exchange capacity of bryophytes: Puustjärvi (1956, 1958)

CHAPTER SEVENTEEN
PROSPECTING
IN
PEATLANDS

A large area of the land in northern latitudes is covered with water-saturated organic terrain known variously as *bog*, *muskeg*, *peatland*, *swamp*, etc. In Canada alone, about 15% of the total area of the country is covered with this type of formation. The peatlands form a dense covering above the bedrock and are a formidable barrier to prospecting for minerals in the substrata below the organic layer. The magnitude of this problem in Canada and in the Soviet Union has been discussed by Hawkes and Salmon (1960) and Shvartsev (1966), respectively. A recent and extensive review by Usik (1969) concerns the use of geobotanical and biogeochemical methods of prospecting in peatlands and is the source of much of the basic geobotanical information contained in this chapter.

Peatlands can be classified into two main groups depending on their hydrological condition. The first of these groups is said to be *minerotrophic* in origin and derives its water from drainage or percolation through rock, till, soil, etc. *Ombrotrophic* peatlands receive water solely from local precipitation and are deficient in mineral-influenced drainage waters. Biogeochemical or geochemical prospecting methods will tend to be more successful in minerotrophic peatlands, as movement of water will increase the chance of detecting mineralization. Even in minerotrophic formations, however, there are still many problems which render prospecting difficult. The first of these is the ready tendency of peat to complex with a number of elements, particularly copper, iron, manganese, uranium, and zinc. The result of this chelatory effect is to increase background levels for a number of elements so that anomalies are harder to detect. Movement of water in the bog can also have the effect of producing a displaced dispersion halo away from the original source of mineralization.

In the broadest definition of the term, biogeochemical prospecting includes the analysis of all organic material; therefore, analysis of peat can be considered a part of the technique and will be discussed briefly in this chapter. Extensive pioneering work on the relationship between peat or peat vegetation and the environment has been carried out by Salmi (1949, 1950, 1955, 1956, 1958, 1959, 1963).

17-1   FACTORS AFFECTING ADSORPTION OF ELEMENTS BY PEAT

Like most humic material, peat has an extraordinary capacity to adsorb cations. Szalay (1954, 1958) has undertaken extensive investigations on this problem, principally in connection with the adsorption of uranium. Adsorption of this and other elements is highly pH dependent and is at a maximum in the pH range 4–10. Below a value of pH 4, adsorption decreases rapidly due to the excess of hydrogen ions which are preferentially adsorbed by the peat. In the optimum pH range, the *distribution coefficient* (concentration in the peat divided by the concentration in the aqueous phase) for the $(UO_2)^{2+}$ ion is 10,000 (fully saturated peat can contain nearly 10% uranium on a dry-weight basis) and for Th^{4+} is 20,000 (Szalay, 1958). In general, distribution coefficients increase strongly with the valence and atomic weight of the cation. The various cations compete with each other and those of higher valence and atomic weight displace others with lower values of these two variables.

The principal agents for fixation of elements by peats are the humic acids. Szalay (1958) has shown that the adsorption of ions decreases to a negligible amount if the humic acids are removed or neutralized. Chelation of copper to organic matter has been described by Manskaya et al. (1958, 1960). Pratt et al. (1967) reported on the chelation of copper and nickel by the same material. The association of molybdenum and uranium with humic acids has been reported by Szalay and Szilagyi (1967) whereas for uranium alone, this association has been reported by Kranz (1968), Manskaya et al. (1956), and Titaeva (1967).

There is a close association between the adsorption of elements by peat and by bryophytes (see Chapter 16) since one of the major sources of many peat bogs is sphagnum moss whose cation-exchange capacity has been studied extensively by Puustjärvi (1956, 1958).

Since adsorption of cations is highly pH dependent, pH measurements should always be made whenever peat samples are taken. In the critical range pH 2–4, even small variations in the value can greatly affect the adsorptive properties of the peat.

Elemental distributions in peat profiles are usually not uniform. Work by Hvatum (1964) and Mitchell (1954) has shown that lead, molybdenum, and zinc tend to concentrate in the upper layers of peat profiles; whereas copper, cobalt, and manganese are sometimes concentrated at the surface and bottom of peat bogs. It is not known to what extent surface enrichment is due to biogenic migration (see biogeochemical cycle, Chapter 10), but the general distribution of elements within the peat profile is probably a function of a large number of factors including Eh and pH.

A partial solution to controlling some of the factors affecting elemental distributions in peat profiles has been proposed by Shvartsev (1966) who suggested that concentration ratios of pairs of elements should be taken as a measure of the presence or absence of mineralization in the area. This is analogous to the classical work of Warren and his co-workers in the use of element ratios in plant material.

17-2 MINERAL BOGS AND PROSPECTING BY ANALYSIS OF PEAT

The concentrations of elements in peat bogs sometimes become so high that the peat itself becomes an economic source of minerals. Records of such mining ventures extend well back to the early nineteenth century, where references can be found to "copper bogs" in Ireland (Townsend, 1824) and Wales (Henwood, 1857). Henwood reported that the peat fields of Merioneth at one time gave an annual yield of copper worth $100,000. Unfortunately, extraction of elements from many bogs presents as yet unresolved economic and technological problems (Bowes et al., 1957).

Elements found in mineral bogs are few in number, but these are all of economic importance. Table 17-1 lists some of the better known mineral bogs in various parts of the world and gives the appropriate references. Some of the bogs are at present mined for their mineral content. One of the best known of these is at Otanmäki in Finland (Salmi, 1955). The ore field is mined from dry areas of the region but extends beneath several large bogs. Investigations by Salmi showed anomalous amounts of iron, titanium, and vanadium in the peat material at points corresponding to mineralized outcrops in the substrate. There was also a correlation between the pH of the peat and the occurrence of ore bodies in the bedrock. Figure 17-1 shows this correlation for the Malmisuo Bog in northern Finland.

The association of relatively high pH values (6–7) with mineralization at Otanmäki is probably due to the presence of lenses of limestone which are commonly associated with zinc in Finland (Salmi, 1958).

TABLE 17-1 Occurrences of Mineral Bogs in Various Parts of the World

Element	Locality	References
Copper	Canada	Fraser (1961a, 1961b)
	Finland	Salmi (1955, 1956)
	Ireland	Townsend (1824)
	Soviet Union	Manskaya et al. (1960)
	United States	Eckel et al. (1949); Forrester (1942); Lovering (1927)
	Wales	Henwood (1857)
Iron	Canada	Hawkes and Salmon (1960)
	Finland	Salmi (1955)
Manganese	Canada	Brown (1964); Uglow (1920)
Uranium	Australia	Debnam (1955)
	Soviet Union	Kochenov et al. (1965); Lisitsin et al. (1967); Moiseenko (1959)
	Sweden	Armands and Landergren (1960)
	United States	Bowes et al. (1957)
Vanadium	Finland	Salmi (1955)
Zinc	Canada	Boyle and Cragg (1957); Ermengen (1957)
	Finland	Salmi (1956)
	United States	Cannon (1955)

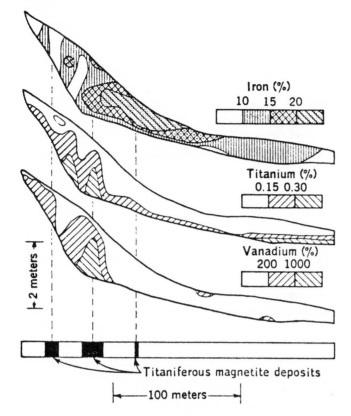

FIGURE 17-1 Section through the Malmisuo Bog, Finland. It shows the relation of iron, titanium, and vanadium in peat to the underlying ore. From Hawkes and Webb (1962) after Salmi (1955).

Mineralization can be detected in peat bogs even when the ore body is covered with an overburden of sand or till up to 40 ft in depth (Usik, 1969).

Procedures in prospecting with peat involve sampling with a bog drill and immediate measurement of pH and Eh. The material is then dried, ashed at 450°C, and analyzed for its elemental content by some suitable method such as atomic absorption or emission spectrography.

Usik (1969) has recommended that detailed local investigation on peat bogs should involve the following five basic steps.

1. Studies of physical conditions such as groundwater flow, diffusion of elements, etc.
2. Studies of elemental distributions in vertical profiles including the underlying clays and tills
3. Studies of the areal and vertical distribution of pH, Eh, and elemental concentrations

   4. Studies of the nature of chelation of elements to organic material
   5. Botanical and ecological studies

The first four of these factors have already been considered to some extent, the botanical aspect will now be considered.

17-3   THE GEOBOTANY OF PEATLANDS

A substantial part of the science of indicator geobotany (see Chapters 3 and 4) has been devoted to a study of the vegetation of bogs. In the well-known work by Chikishev (1965), 4 of the 41 contributed papers are exclusively devoted to peatlands.

In indicator geobotany in peatlands, two main types of indicator are studied. The first of these is a group of *direct indicators* whose presence alone is a measure of the existence of mineralization or at least of a high content of a particular element in the peat. Among direct indicators may be cited copper mosses (see Chapter 4) such as *Pohlia nutans* whose presence indicates a high copper content in the bog (Fraser, 1961a).

Indirect geobotanical indicators are useful in defining the physical, hydrological, or chemical conditions of peatlands from which mineralization might be inferred directly or discovered later as a result of biogeochemical investigations carried out on the basis of the knowledge gained of the environmental conditions.

Classification of the various types of peat bogs can be effected by a study of their vegetation. For example, minerotrophic conditions are characterized by a fen vegetation containing basiphilous or calcareous plants. The degree of minerotrophy can be calculated from the relative abundance of the various plant species.

Ombrotrophic peats tend to be poor in nutrients and have a low pH. This results in plant associations typical of this type of formation, with dominance of such species as *Picea mariana* (black spruce) and *Ledum palustre* or *Ledum groenlandicum* (both sometimes known as Labrador tea).

Classification of peats according to their vegetation cover has been undertaken by numerous workers (Sjörs, 1950a, 1961, 1963; Gorham, 1956; Ritchie, 1960, 1962; Heinselman, 1963, 1965; Sparling, 1966; Bay, 1967). The vegetation cover is also an indication of the pH of the peat and this factor has been discussed by Bay (1967), Pierce (1953), and Sjörs (1950b). Examination of the plant cover can also afford some indication of the geologic substrate beneath the bog, provided that the peat layer is not too thick.

Aerial photography with black and white, color, or infrared film is finding widespread use in the evaluation of organic terrain. Aerial photographs are particularly useful because of the inaccessible nature of peatlands. Work of this type has been carried out by Radforth (1955, 1958) and by Radforth and Usik (1964).

17-4   PROSPECTING IN PEATLANDS BY ANALYSIS OF VEGETATION

Records of biogeochemical prospecting in peatlands are very meager, due no doubt to the inherent difficulties of this type of work in such an environment. Warren and Delavault (1955c) have reported that biogeochemical methods can be employed in northern latitudes by use of the common bog plants, *Picea mariana* and *Ledum groenlandicum*. Poskotin and Lyubimova (1963) have reported that the birch and pine are effective in delineating copper anomalies in peat bogs.

The most extensive investigations on the use of bog plants for biogeochemical prospecting seem to have been carried out in Finland (Rankama, 1941; Marmo, 1953; Salmi, 1956). The work of Marmo and Salmi was based on a single species: the ubiquitous *Ledum palustre*.

Marmo (1953) carried out his investigations in the Rautio region (Northern Finland) which is covered with bog to a depth of 9 to 20 ft. Good agreement was obtained between the molybdenum content of the leaf ash and the known occurrence of this element in the bedrock. Copper anomalies were also detected by analysis of the leaf ash of the same species.

Extensive biogeochemical work by Salmi (1956) on *L. palustre* growing on a mineralized bog in Northern Finland confirmed that the biogeochemical method may be applied in peatlands. The area of study was in the Vihanti region where the commonest rocks were mica-schist, quartzite, amphibolite, and granite. The country rock was overlain by 3–30 ft of till and peat. The ore zone contained mainly iron and zinc with abundant pyrite and sphalerite. Chalcopyrite and galena were also present as auxiliary minerals.

Peat samples and specimens of leaves and twigs were collected at intervals along an east–west traverse of the bog. The concentrations of copper, lead, and zinc in all the vegetation samples was significantly higher in the western part of the bog which was known to overlay mineralization.

Salmi determined that sampling of vegetation was most useful as a prospecting technique in bog regions with a thin layer of peat and an overabundance of water. When the layer was thick and water not excessive, sampling of the peat itself was a better indication of mineralization below the overburden.

SELECTED REFERENCES
General review: Usik (1969)
Geobotany of peatlands: Chikishev (1965)
Biogeochemistry of peatlands: Salmi (1956)

CHAPTER EIGHTEEN
A CASE HISTORY
OF
INTEGRATED WORK
IN NEW ZEALAND

Biogeochemical prospecting is useful as an adjunct to other exploration methods and serves to complete the whole picture of elemental distributions within a particular mineralized area or biogeochemical province. Its greatest value probably lies not so much in its use as an exploration method per se but rather as part of an integrated program of geophysical, geochemical, and biogeochemical exploration techniques.

The following account concerns an example of integrated work undertaken in New Zealand in 1965–1966 and shows how a copper–molybdenum anomaly was delineated by geochemistry and confirmed and extended by biogeochemistry. Geophysical work of an experimental nature was also carried out. Fuller details of the biogeochemical work have been reported by Brooks and Lyon (1966), Lyon (1969), and Lyon and Brooks (1969).

The anomaly is situated at Copperstain Creek, a tributary of the Pariwhakaoho River about eight miles from Takaka in the northwest of the South Island, New Zealand.

18-1   CLIMATE, TOPOGRAPHY, AND VEGETATION
OF THE MINERALIZED AREA

The climate of the Copperstain Creek area is generally mild, with an annual rainfall of 80–100 inches distributed evenly throughout the year.

The terrain is extremely steep and dissected with numerous small streams. Figure 18-1 shows a geologic map of the region. The elevation at the mouth of Copperstain Creek is about 600 ft and rises steadily to about 1000 ft at its junction with Moly Creek. The top of the ridge separating Copperstain Creek from Mineral Creek to the southwest is at an elevation of 1600 ft.

The vegetation belongs to the "lower beech forest" zone with types characteristic of the North Island rather than the South Island. The tree layer is dominated by *Nothofagus fusca* and *Weinmannia racemosa*. Broadleaf species such as *Myrsine*

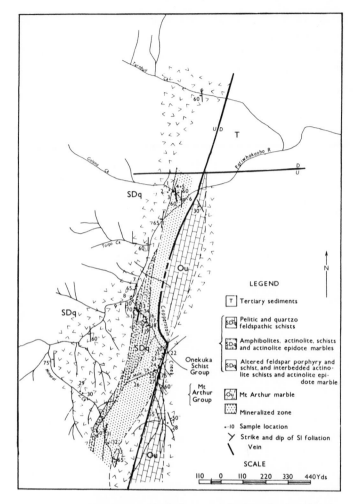

FIGURE 18-1   Geologic map of mineralization in the
Takaka area, South Island, New Zealand.   From Ellis
(1966).   By courtesy of Chemistry Division D.S.I.R.,
New Zealand.

salicina, Olearia rani, and *Quintinia acutifolia* are small trees common beneath the
forest canopy.

18-2   THE GEOLOGY OF THE COPPERSTAIN CREEK AREA

The mineralization occurs in a faulted zone with Mt. Arthur marble on
the east of a north-south fault and a complex and mineralized sequence of impure
marbles, schists, and amphibolites on the west intruded by several small sills and
a small cupola of granite.   The mineralization extends for about 1500 ft in a
300–400 ft band rich in pyrite.

The rocks have been subjected to more than one period of faulting and have undergone varying degrees of metamorphism and metasomatism.

The soils of the area are skeletal and poorly formed due to the broken nature of the terrain where erosion can readily occur. The geologic features of the area are shown in Figure 18-1.

18-3   GEOPHYSICAL INVESTIGATIONS

In 1965 geophysical investigations were undertaken by Lime and Marble Ltd. of Port Mapua, Nelson. Readings were carried out with a portable continuous-reading "Fluxgate Magnetometer" at intervals of 100–200 ft along the main access road up Copperstain Creek, across and into the headwaters of Mineral Creek, and then back along the western ridge of Copperstain Creek to its confluence with the Pariwhakaoho River. This survey, which was aimed at delineating zones of intense alteration (associated with magnetite), showed that to be effective, readings on a 50-foot grid would have had to have been employed. This was not attempted.

Further geophysical work involved the use of an ABEM Electromagnetic Gun based on the "Slingram" method but again was unsuccessful due to the steep topography.

Geophysical work with AFMAG and EM equipment was ultimately tested in the area. Although EM gave no response because of the disseminated nature of the mineralization, AFMAG appeared to delineate the boundary of the sulfide bodies.

18-4   HYDROGEOCHEMICAL INVESTIGATIONS

A program of stream-water sampling in the area was undertaken by the Chemistry Division of the New Zealand Department of Scientific and Industrial Research (Ellis, 1966). Total heavy metals (copper + zinc) were determined colorimetrically *in situ* by use of the dithizone method. The data are shown in Figure 18-2, from which it can be seen that anomalous amounts of heavy metals were found near the junction of Copperstain and Moly Creeks.

18-5   GEOCHEMICAL INVESTIGATIONS

Stream sediments were analyzed by Chemistry Division DSIR (Ellis, 1966). Cold-extractable copper (solvent—10% ammonium citrate/hydroxylamine buffer of pH 5) was determined by atomic-absorption spectrophotometry. The results are shown in Figure 18-2. The same figure gives values for total molybdenum in the stream sediments and shows highly anomalous values near the junction of Copperstain and Moly creeks.

Soil samples taken from a 100-foot rectangular grid were analyzed for copper, molybdenum, and other elements both of D.S.I.R. and by the author's own research group at Massey University. Both sets of data indicated significant copper and molybdenum anomalies.

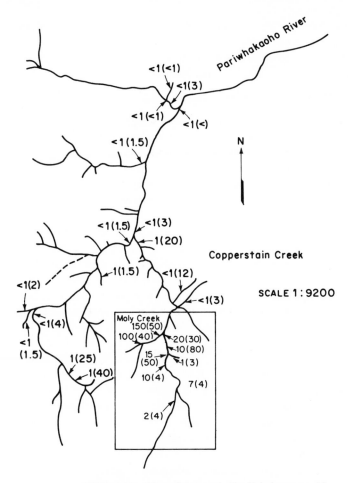

FIGURE 18-2   Map of streams in the Takaka areas, New
Zealand, showing values for molybdenum concentrations
(ppm) in stream sediments (not in parentheses) and for
total heavy metals (ppb) in stream waters (values in
parentheses). Area in box is that covered by Figure 13-4.
From Ellis (1966). By courtesy of Chemistry Division
D.S.I.R., New Zealand.

18-6   BIOGEOCHEMICAL INVESTIGATIONS

PRELIMINARY WORK   In preliminary work undertaken in 1965 by the
author's group (Brooks and Lyon, 1966), four species of vegetation: *Myrsine
salicina, Olearia rani, Quintinia acutifolia*, and *Weinmannia racemosa*, were sampled
from various sites over a traverse of the mineralized area in the vicinity of the junction of Copperstain and Moly creeks. All four plants gave a response to a high
molybdenum concentration in the soils. These species had been selected mainly
because of their size and frequency of occurrence in the area. *W. racemosa* was
later discarded as a suitable species for prospecting because it was so large that

sampling problems arose from the inaccessibility of its leaves and twigs. Copper and molybdenum were analyzed in these organs for the remaining three species. Molybdenum concentrations were in all cases appreciably higher in leaves than in twigs (expressed on an ash-weight basis); copper showed a different trend, with slightly higher concentrations in twigs.

AN ORIENTATION SURVEY   The three species selected for further study were sampled on a rectangular grid basis at 100-foot intervals over the whole of the exploration area. Leaves rather than twigs were sampled since the preliminary work indicated that these might be more suitable for the purpose. Moreover, *M. salicina* with its thick fleshy stem did not readily afford suitable woody material.

Table 18-1 gives the geometric means and concentration ranges for copper and molybdenum in each of the three species studied.

In Figure 18-3, the plant–soil relationship for molybdenum concentrations in the leaf ash of *O. rani* is shown on logarithmic coordinates. This species apparently shows a partial exclusion mechanism for concentrations up to 80 ppm in the soil. Above this threshold, there is a greatly increased accumulation of molybdenum and the amount in the plant apparently tends to infinity for high concentrations of this element in the soil. The highest concentration found in any specimen of leaf ash was however 1600 ppm (not shown in Figure 18-3) and this probably represents the true limiting value.

STATISTICAL INTERPRETATION OF THE DATA OF THE ORIENTATION SURVEY
The data were processed statistically with the aid of a computer in order to obtain values for the correlation coefficients (r) for plant–soil data for copper and molybdenum. The statistical data are presented in Table 18-2. Concentrations were expressed on a logarithmic basis since other statistical work had indicated that elemental distributions were closest to log-normality in character.

In addition to values of r for plant–soil data, calculations were also made for the interelemental copper–molybdenum relationship in plants and soils considered separately, in order to establish whether there was any evidence for mutual antagonism or stimulation (see Chapter 11).

TABLE 18-1   Copper and Molybdenum Concentrations in Leaf Ash and Soils from Copperstain Creek, New Zealand

Material	No. samples	Element	Geometric mean	Range
Myrsine salicina	42	Copper	86	1–210
	42	Molybdenum	8	1–80
Olearia rani	45	Copper	207	25–840
	45	Molybdenum	18	1–260
Quintinia acutifolia	44	Copper	142	32–380
	44	Molybdenum	12	1–250
Soils	64	Copper	104	32–380
	64	Molybdenum	72	9–215

SOURCE: Lyon (1969).

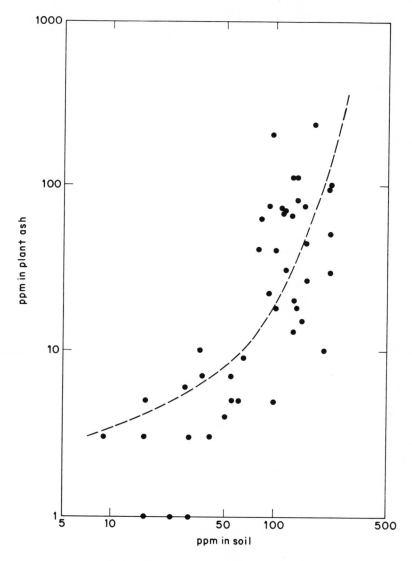

FIGURE 18-3 Plant–soil relationship for molybdenum in the ash of leaves of *Olearia rani*. From Lyon (1969).

The data in Table 18-2 showed that there was a very highly significant correlation between the molybdenum content of the leaf ash of *O. rani* and the concentration of this element in the soil.   There was also a highly significant relationship for *M. salicina* and *Q. acutifolia*, though in both cases, levels of significance were much lower than for *O. rani*.

The data for copper were disappointing. Plant–soil correlations for this element were in the main nonsignificant; indeed, in the case of *Q. acutifolia* there was apparently a significant inverse relationship. This was probably due to the fact that

TABLE 18-2 Correlation Coefficients (r) for Logarithms of Concentrations (ppm) of Copper and Molybdenum in 45 Samples of Leaf Ash and Associated Soils from Copperstain Creek, New Zealand

Relationship	Element	*M. salicina* r	Signif-icance	*O. rani* r	Signif-icance	*Q. acutifolia* r	Signif-icance	Soils r	Signif-icance
Plant × soil	Copper	−0.15	NS	−0.26	NS	−0.42	S*	—	—
Plant × soil	Molybdenum	0.46	S*	0.77	S**	0.41	S*	—	—
Interelement	Copper × molybdenum	0.03	NS	0.33	S	0.04	NS	0.26	S

S** = significant at 0.1% level.
 S* = significant at 1% level.
  S = significant at 5% level.
 NS = not significant.

SOURCE: Lyon (1969).

copper is an essential element (see Chapter 14) so that relative uptakes decrease as the copper concentration in the soil increases. There was fortunately no evidence for a mutual antagonism between copper and molybdenum, as shown from values of r for the interelemental relationship in the plant ash of all three species studied.

Cumulative frequency plots (see Chapter 13) were obtained for molybdenum in the leaf ash of *O. rani* and in soils. The data are shown in Figure 13-3, which shows that values of 125 ppm and 60 ppm represent threshold levels for molybdenum in soils and leaf ash, respectively.

Figure 13-4 shows isoconcentration contours for areas of anomalous amounts of molybdenum in the leaf ash of *O. rani* and in soils; the above threshold levels were used as criteria of mineralization. The same figure also shows concentration contours for very highly mineralized areas (greater than 125 ppm in plant ash and greater than 60 ppm in soils). Even the lower threshold levels seem considerably above normal background concentrations, but the high value is probably due to extensive overlap between mineralized and nonmineralized distributions. The values obtained from cumulative frequency curves are *objective* in nature and do not depend on the personal bias of the field worker.

A FOLLOW-UP ORIENTATION SURVEY FOR MOLYBDENUM In view of the apparent success of *O. rani* as a prospecting agent for molybdenum, further work was carried out on this species in which a further twenty-six specimens of plants, together with associated soils, were collected at various points on and off the anomaly. Analyses for copper and molybdenum were carried out on flowers, leaves, old wood, and 1-year-old twigs (Lyon and Brooks, 1969). The results, shown in Table 18-3, were calculated on both an ash-weight and dry-weight basis. Values of r were computed for various pairs of variables. The statistical data (see Table 18-4) show that the molybdenum content of leaf ash of *O. rani* remained the best criterion of anomalous amounts of this element in the soil. Table 18-4 also shows that there was no difference between molybdenum results expressed on an ash-weight and a dry-weight basis. Equally good results would have been obtained

TABLE 18-3   Copper and Molybdenum Concentrations in the Ash of Various Organs of 26 Specimens of *Olearia rani* and Associated Soils from Copperstain Creek, New Zealand

Material	Element	Geometric mean (ppm)	Range (ppm)
Flowers	Copper	206	140–360
	Molybdenum	24	4–72
Leaves	Copper	142	60–310
	Molybdenum	59	10–1600
Old wood	Copper	231	70–480
	Molybdenum	24	3–210
One-year-old twigs	Copper	303	60–840
	Molybdenum	19	3–145
Soils	Copper	180	100–410
	Molybdenum	197	58–520

SOURCE: Lyon (1969).

TABLE 18-4   Correlation Coefficients (r) for Logarithms of Concentrations (ppm) of Copper and Molybdenum in Organs of 26 Specimens of *Olearia rani* and Associated Soils from Copperstain Creek, New Zealand

		Relationship					
		Plant × soil				Interelemental	
		Copper		Molybdenum		Copper × molybdenum	
Material	State	r	Significance	r	Significance	r	Significance
Flowers	Ashed	0.27	NS	0.37	NS	0.40	S
	Dried	0.19	NS	0.32	NS	—	—
Leaves	Ashed	0.08	NS	0.48	S	0.27	NS
	Dried	0.16	NS	0.48	S	—	—
Old wood	Ashed	0.03	NS	0.29	NS	0.54	S*
	Dried	−0.02	NS	0.29	NS	—	—
1-year-old twigs	Ashed	0.27	NS	0.50	S	0.15	NS
	Dried	0.36	NS	0.45	S	—	—
Soils	Ignited	—	—	—	—	0.38	S

N.B. Significance symbols as in Table 18-3.

SOURCE: Lyon (1969).

by analysis of 1-year-old twigs. The significance of r (0.48) for only 26 samples was less than for the previous survey, possibly because the samples were selected in a random manner rather than on the basis of a rectangular grid. However, when the two sets of data were combined (71 samples), the overall value of r for the plant–soil relationship was 0.74, which is even more significant than that obtained previously (0.77) for only 45 specimens.

In general, there was a slightly better plant–soil correlation for elemental values expressed on an ash-weight rather than dry-weight basis; this confirms similar findings by Warren et al. (1955).

Values of *r* for the copper–molybdenum relationship in all plant organs sampled showed no evidence for any mutual antagonism. It is clear also from Table 18-4 that old wood and flowers of *O. rani* are not suitable for prospecting purposes.

18-7   DRILLING PROGRAM AND ASSESSMENT OF THE ANOMALY

Following the exploration work described above, diamond drilling was carried out by Lime and Marble Ltd. and was completed in November 1966. This work was undertaken in the area near the junction of Copperstain and Moly Creeks, indicated as being anomalous by all the biogeochemical and geochemical data and to some extent by the AFMAG geophysical data as well.

As a result of the drilling program, this company has reported probable reserves of about 10 million tons of ore averaging 7.4% sulfur and 0.15% copper. A further 1.2 million tons is said to average 0.02% copper and 0.05% molybdenum.

18-8   GENERAL CONCLUSIONS

The case history recorded in this chapter is of particular interest for New Zealand insofar as it represents the first attempt at biogeochemical prospecting in that country. The region itself is of interest because it represents an area where the cheaper geophysical methods were only partially effective. Use of EM equipment was ineffective due to the disseminated nature of the ore minerals and also because of the severe topography. Magnetics required such small spacing of stations due to the distribution of magnetite that the cost of such close control in the initial stages of the investigation was not warranted. AFMAG delineated at least the apparent western boundary of the mineralized zone and was not tested in the east. This method would probably have been effective in delineating the mineralized zone in a reconnaissance role. The technique of induced polarization was not tested but would probably have responded well to the environment.

In contrast to geophysics, biogeochemical and geochemical methods were quite effective in this region in delineating anomalies. To assess the overall success of the integrated program, the effectiveness of the biogeochemical and geochemical techniques is summarized in Table 18-5.

Geochemical methods involving the analysis of stream waters would have satis-

TABLE 18-5   A Comparative Assessment of Some Methods of Exploration Used at Copperstain Creek, New Zealand

	Method used			
	Biogeochemical	Geochemical		
	Leaf ash	Sediments	Soils	Water
Target	Ground anomaly	Drainage area	Ground anomaly	Drainage area
Effectiveness for copper	Poor	Good	Good	Good
Effectiveness for molybdenum	Good	Good	Good	Poor

factorily delineated the drainage area containing the copper mineralization but would not have shown the presence of molybdenum.

Analysis of stream sediments was effective in demonstrating the presence of both copper and molybdenum mineralization in the drainage areas concerned.

Soil analyses were apparently successful in delineating both the copper and molybdenum anomalies, whereas the biogeochemical method was successful only for molybdenum.

In considering the case history as a whole, it is not easy to decide what methods were the most useful for prospecting purposes. The area in which drilling was ultimately carried out could have been pinpointed equally well by analysis of plants or soils and it is significant that the position of the target was confirmed by no less than five different procedures.

It may not always be necessary to apply so many methods to a new exploration area, but there can be little doubt that the confidence with which drilling can be undertaken increases as more techniques are used to delineate the target area. With the current high cost of drilling operations, this is a consideration that cannot be ignored.

SELECTED REFERENCES

Biogeochemical prospecting at Copperstain Creek: Brooks and Lyon (1966); Lyon and Brooks (1969)

Geochemical prospecting at Copperstain Creek: Ellis (1966)

CHAPTER NINETEEN
AN
ASSESSMENT OF
BIOGEOCHEMICAL
PROSPECTING
METHODS

The true effectiveness of biogeochemical methods of prospecting is hard to establish. Although the literature records many biogeochemical surveys over known mineralized areas or regions later evaluated by some other method, in prospecting, the biogeochemical approach is seldom used by itself. The many references to biogeochemical prospecting in the Soviet literature do not often discuss the effectiveness of the method.

In the author's own experience in New Zealand, every single anomaly investigated would certainly have been discovered if biogeochemical methods alone had been used. Warren and his co-workers, in their pioneering work in British Columbia, reached the same conclusion (Warren and Delavault, 1950b).

Comments in Chapter 7 concerning the reluctance of many field workers to extend geobotanical work beyond areas of known mineralization are certainly valid for biogeochemical work also. In cases where the method was used as the *initial* exploration procedure, subsequent investigations by other techniques have usually confirmed the value of the biogeochemical operations. A few examples of such cases are given in the next section.

19-1   SOME SUCCESSFUL USES OF BIOGEOCHEMISTRY IN MINERAL EXPLORATION

Biogeochemical methods have been successfully used in the search for uranium in the Colorado Plateau Region of the United States. Nearly 11,000 tree samples were collected for uranium analysis in the Grants District, New Mexico and at Elk Ridge, Utah. Cannon (1960a) reported that from work by F. J. Kleinhampl at Elk Ridge, radioactive sandstone was found in 80% of the "plant" anomalies detected. Of 108 anomalous localities delineated by plant analysis, 55 may have contained ore.

In the Grants District, "plant anomaly" maps compiled from data on the uranium content of junipers and pinyon pines were found to agree well with the results of drillings at a later date (Cannon, 1960a). Cannon has also reported that a program involving the sampling of *Atriplex confertifolia* on dune sands of the Carrizo Mountain area was not successful because of the shallow nature of the roots system of this species.

The effective depth of conifers for detecting uranium was about 40 ft at Elk Ridge and as much as 70 ft in the Circle Cliffs region (Kleinhampl and Koteff, 1960).

In a later project in the Yellow Cat area, Grand County, Utah, 400 samples of trees and shrubs were analyzed for uranium in areas of potential mineralization (Cannon, 1964). Unfortunately the drilling program was not sufficiently extensive to evaluate fully the potential of the method in this area. The limited drilling program undertaken on the basis of analysis of *A. confertifolia* did reveal the existence of two shallow uranium ore bodies. Both this species and the juniper were effective in delineating ore bodies at a depth of up to 20 ft. Below this depth, only the juniper was effective.

Keith (1968) has reported that biogeochemical methods involving the analysis of material from elms, maples, and oaks of the Upper Mississippi Valley District were effective in detecting lead and zinc ore bodies below an overburden of loess under circumstances where no anomalies were apparent in the soils. In other parts of the area, where the loess was lacking, soil sampling was the more effective procedure.

Riddell (1952) successfully carried out biogeochemical prospecting at Gaspé-North County, Quebec, by analysis of balsam twigs for copper. His work led to the discovery of several previously undiscovered ore bodies.

The success achieved by the considerable volume of Russian work is more difficult to assess; but Malyuga (1964) has reported at least six cases of the discovery of ore deposits in the Soviet Union by biogeochemical methods alone.

19-2   SOME OBSERVATIONS ON THE ECONOMICS OF BIOGEOCHEMICAL PROSPECTING

There has already been some discussion (Chapter 15) on the economic factors involved in the analysis of plant material compared with rocks or soils. Although there is apparently an extra step involving ignition of the sample before analysis, this is offset by the greater ease with which plant ash (compared with soils) can be dissolved. In general, therefore, the cost of analyzing plant material should approximate that for rocks and soils.

In comparing the economics of geochemical and biogeochemical methods of prospecting, the main factors will be therefore the difference in the cost of collection of the samples. Whether soil sampling or vegetation sampling is speedier depends very largely on the terrain. In areas of thick forest cover with a deep layer of humus, the true soil is often difficult to find. This is especially true in some parts of New Zealand where the humic layer can be several feet thick and covered in turn with a dense mat of forest litter and intertwining roots. Under such conditions, plant sampling is appreciably quicker and less tiring. Under more arid conditions, where the soil is immediately available, sampling of vegetation will be more time consuming.

Figure 19-1 lists economic and other factors concerning the selection of exploration methods and places biogeochemistry and geobotany in perspective with other geochemical exploration methods.

19-3   ADVANTAGES AND DISADVANTAGES OF
        BIOGEOCHEMICAL PROSPECTING METHODS

Some of the advantages and disadvantages of biogeochemical methods of prospecting are listed below.

ADVANTAGES

1. Large trees can sometimes penetrate through a thick overburden to give evidence of mineralization at depth. An effective depth of at least 100 ft has been reported for the juniper in prospecting for uranium in the Colorado Plateau.
2. Plants with extensive root systems can effectively sample a greater soil volume than the size of samples normally collected in soil surveys.
3. In densely overgrown areas with a deep humus layer in the soil, plant sampling is easier and quicker than soil sampling.
4. Plant samples are usually appreciably lighter than soil samples and this factor is useful in field work.
5. Chemical analysis of plant ash is sometimes more free from interference problems since the concentrations of elements such as iron which frequently cause problems in chemical procedures are substantially less in plant ash than in rocks or soils.
6. In cases where effective soil sampling depends on the selection of a particular horizon, plant sampling can be much simpler since it only depends on the selection of a particular easily available and identifiable plant organ.
7. In cases where the soil is either almost nonexistent or complicated by the existence of such features as a siliceous hardpan, biogeochemical prospecting can be more advantageous than soil sampling.
8. The method is particularly useful in areas of permafrost or in cold regions frozen for a large part of the year.

DISADVANTAGES

1. Plant sampling demands more skill and experience from the field worker, particularly as the question of recognition of species arises.
2. Unknown factors, such as pH, drainage, aspect, and age of organ, can greatly affect the reliability of the method unless they can be controlled or allowed for.
3. The method depends on an even distribution of the selected species in the exploration area or at least the availability of more than one species for prospecting.

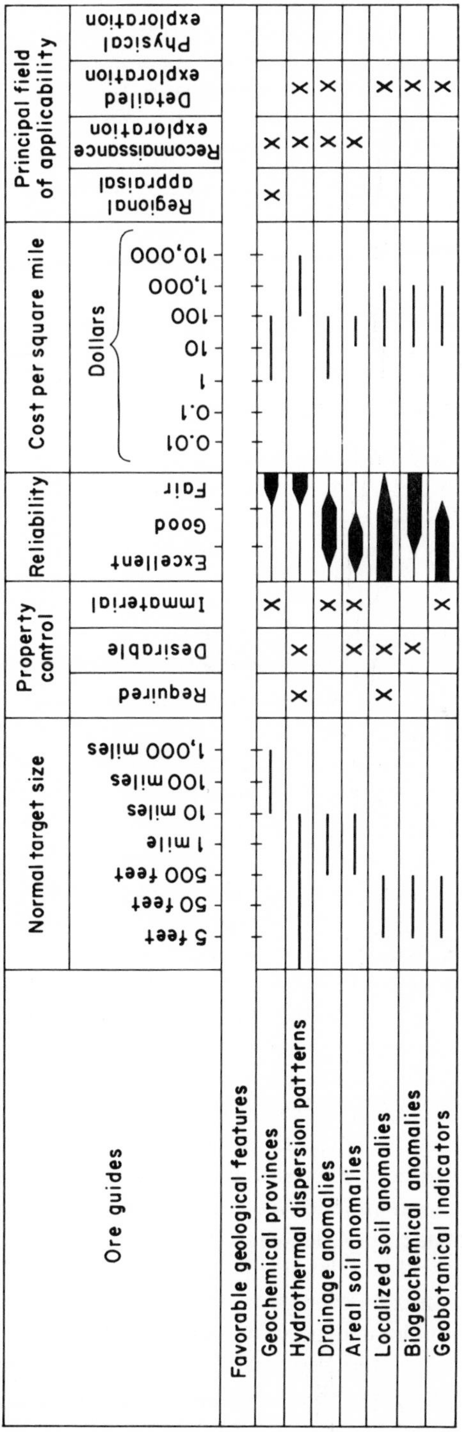

Ore guides	Normal target size (5 feet – 1,000 miles)	Property control — Required	Property control — Desirable	Property control — Immaterial	Reliability (Fair / Good / Excellent)	Cost per square mile, Dollars (0.01 – 10,000)	Regional appraisal	Reconnaissance exploration	Detailed exploration	Physical exploration
Favorable geological features										
Geochemical provinces				X	Fair		X			
Hydrothermal dispersion patterns			X		Fair			X	X	
Drainage anomalies				X	Good–Excellent			X	X	
Areal soil anomalies			X		Good–Excellent			X		
Localized soil anomalies		X			Excellent–Good				X	
Biogeochemical anomalies			X		Fair–Good				X	
Geobotanical indicators				X	Good				X	

FIGURE 19-1   Comparison of factors in geochemical, biogeochemical, and geobotanical methods of prospecting. After Hawkes and Webb (1962).

4. Biogeochemical prospecting methods usually require an orientation survey over a known anomaly before they can be used with confidence in a new area. The same is admittedly true for soil sampling where a decision has to be made as to which horizon should be sampled. Such soil surveys are, however, simpler than those involving vegetation.
5. Sometimes the method can only be applied at certain seasons of the year (e.g., during the growing period, if sampling of deciduous leaves is involved).
6. Biogeochemical methods are not invariably applicable whereas soil sampling methods usually are.
7. Plant samples are much more subject to contamination than soils.
8. Elements essential in plant nutrition can sometimes cause difficulties in biochemical prospecting (see Chapter 14).

The above listing of the advantages and disadvantages of the biogeochemical method is far from complete but it may be concluded that the greatest single advantage of this technique is its penetrating power and that its greatest disadvantage is its variability caused by factors difficult to control.

**19-4   CONSIDERATIONS IN THE CHOICE
OF THE BIOGEOCHEMICAL METHOD**

Whether or not the biogeochemical method is to be employed in a particular area depends on a number of factors. In areas of permafrost or where the soil is shallow or virtually nonexistent, the usefulness of vegetation sampling is obvious. In other cases, however, the need for the method may not be as great as for the simpler procedure of soil sampling.

If biogeochemical methods of prospecting are to be used, there are a few conditions that must be satisfied before the attempt is made. The first of these is that suitable, well-distributed species must be present in the exploration area. The easiest way of making an assessment of this nature is to use infrared color photography. This film gives such good differentiation of vegetation species that distributions can be determined in a matter of minutes instead of days or weeks as would be the case if a ground survey were used to establish distribution patterns of vegetation. This technique is discussed in Chapter 6.

Another condition that should be satisfied before biogeochemical prospecting is undertaken is that the vegetation selected should not only be of widespread distribution but should also consist of suitable species. Suitable species would be plants with deep root systems and yet with a size sufficiently small to render sampling of twigs or foliage a practicable proposition.

In practice, few areas will be completely suitable or unsuitable for biogeochemical prospecting. The ultimate decision on whether or not to use the method will depend on the factors enumerated above as well as economic considerations. If, in addition, an ample pool of knowledge is available for the regional response of vegetation to mineralization, as for example in British Columbia or New Zealand, there will be further incentive to carry out this type of prospecting.

Biogeochemical methods should preferably be carried out in conjunction with other methods, if economic factors make this permissible; additional information can never be otherwise than beneficial. When biogeochemistry confirms anomalies detected by other methods, this is useful, but where there is a conflict of results, investigation of the reasons for such discrepancies can often lead to fresh discoveries of direct use in prospecting. For example, an anomaly in plants not confirmed by the soils in an area of steep terrain might indicate that soluble salts had been leached from another anomaly upslope from the test area.

19-5   FUTURE DEVELOPMENTS IN BIOGEOCHEMICAL PROSPECTING
The application of aerial photography in geobotanical prospecting is obvious, but the use of aerial methods in biogeochemical work is much more difficult to envisage. There is, however, a way in which aerial methods might be applied to advantage in thickly forested regions of the world where access is difficult. This procedure would be actual sampling of vegetation from the air by means of an automatic harvester operated from a helicopter. If such a device could be be developed, biogeochemical prospecting would have a great advantage over soil or stream-sediment sampling in areas of broken terrain. In some parts of the world, particularly in New Zealand and New Guinea, the native bush is so thick and dense that a major cost of any ground operation comprises cutting of trails and access routes.

An Australian exploration company working in New Guinea has attempted to sample stream sediments from the air by means of a tubular missile dropped from a helicopter into stream beds and retrieved with sediments trapped within the tube. Unfortunately, success has been meager due to the relatively small amount of suitable sediment in the streams concerned.

In New Zealand, an attempt has been made to devise a foliage reaper which would cut vegetation by means of a rotating cutter approaching from the side of the tree and suspended from a helicopter operating at tree-top height. Unfortunately, practical problems have been encountered in making the operation sufficiently secure to comply with air safety regulations since there is a tendency under the present arrangement for the tail rotor of the aircraft to become entangled in vegetation. A solution to this problem would be to operate at a higher level to increase the safety of the operation and then to cut vegetation with apparatus which would convey the chopped material into the cockpit by means of a suction device similar to a vacuum cleaner. This, however, presents fresh problems since the higher the aircraft above the sampling point, the greater the power needed to convey vegetation to the cockpit and the less the control over guidance of the cutter. If a satisfactory compromise can be effected between power requirements and safety, it should be possible to sample vegetation from the air with all the advantages that this implies.

One of the fields in which biogeochemical prospecting might be expected to develop during the next decade is in the increasing use of pathfinders as a guide to mineralization. Warren et al. (1964) have reported on the use of arsenic as a pathfinder for sulfide minerals. Some pathfinders have the advantage of greater mobility and hence greater dispersion than the ore elements themselves. Moreover, an element such as arsenic is normally present in such low amounts in vegetation growing over nonmineralized soils that a small concentration of this element derived from mineralization should be readily detectable. The author is of the opinion that further work on pathfinders in vegetation will afford rewarding projects for the future. Work of this nature, involving arsenic, selenium, and tellurium in plants, is at present under way in the author's laboratory.

Other possible trends of biogeochemical prospecting in the future could involve the use of the mass spectrograph to study the origins of sulfates in plant material, since sulfates derived from sulfide mineralization might be expected to have a different isotopic ratio from those derived from other sources.

Direct on site analysis of plant material may be possible in the near future. Already the development of gamma-source portable X-ray fluorescence units is leading to *in situ* analysis of rocks and soils. There is no reason, therefore, when the procedure has been rendered more sensitive by new instrumentation, why it should not be used, for example, to measure the concentrations of elements in tree trunks or other plant organs.

At present, orientation surveys need to be carried out in new areas before biogeochemical prospecting can be applied with complete confidence. If, however, basic research can be undertaken to establish the main parameters affecting elemental uptake by plants, it should be possible to undertake biogeochemical prospecting in a new area merely by making corrections for the various disturbing factors. The use of elemental ratios other than the classical copper/zinc value, should also be investigated in order to reduce the interference effects.

Computer science will continue to play an increasingly important role in the use of biogeochemical methods. Even today, it is possible to feed raw data from an orientation survey into a computer and obtain all the necessary information for a subsequent exploration survey.

The likely trends suggested above are but a few of many. Biogeochemical prospecting has now been in use for over 30 years, but with the increasing availability of new techniques of analysis and computerization, and with the upsurge of new ideas fostered by the technological revolution in the world today, there is no reason why progress in this method of exploration during the next few years should not greatly exceed all that has been achieved in the past.

PART THREE
THE BIOGEOCHEMISTRY OF THE ELEMENTS

CHAPTER TWENTY
THE BIOGEOCHEMISTRY OF THE ELEMENTS

The purpose of this chapter is to summarize available data on the application of geobotany and biogeochemistry to exploration for individual elements. Abundance data are given for elemental concentrations in igneous rocks, soils, and plant ash. Values 'for igneous rocks and soils are based largely on Green (1959) and Vinogradov (1959) as compiled by Andrews-Jones (1968). Data for elemental contents in plant ash and for toxicities to plants are derived mainly from Cannon (1960b) and Bowen (1966). In some cases the values given for concentrations in soils or plant ash are based on data obtained in the author's own laboratory.

The entry for each element also includes a list of references appertaining to biogeochemistry and in most cases to geobotany also. World literature contains a multiplicity of references to the occurrence of trace elements in vegetation and the major proportion of such citations refers to crop and pasture species. In this chapter, references concerning distributions of trace elements in plants are confined almost entirely to those referring to wild species. There are also a few references to the use of analytical chemistry in determining elements in plant ash, but the examples given are restricted to cases of especial interest (e.g., gold in vegetation) and where the aim of the work was to assist in biogeochemical prospecting.

The *Chemical Abstracts* (CA) citation is given for each reference wherever possible; and at the end of each subsection, an abstract of a complete paper is given. These abstracts are usually based on the entry in *Chemical Abstracts*.

The elements discussed are mainly those of current or potential economic significance. The list includes elements (such as selenium) that may be of more importance as pathfinders than as direct indicators of mineralization.

20-1   **ANTIMONY**
Concentration in igneous rocks: 0.3 ppm
Concentration in soils: 0.5 ppm
Concentration in plant ash: 1 ppm

Toxicity to plants: moderate
Geobotanical reference: Flerova and Flerov (1964) (CA 62, 14351f.)
Biogeochemical references: Flerova and Flerov (1964) (CA 62, 14351f.);
   Nikiforov and Fedorchuk (1959) (CA 55, 12173i)

ABSTRACT   Flerova and Flerov (1964). "Experiment on biochemical
and geobotanical prospecting in the Dzungarian Ala Tau." Results are
given of surveys made over polymetallic deposits at Koksu, Eastern
Suuktyube, and Yablonovoe. In each 5 sq mi survey unit, geobotanical, soil, and geologic observations were made and plant and soil
samples were taken. The plants spirea, mint, yarrow, St. John's wort,
and vetch, concentrated large amounts of *antimony*, gallium, and silver
in sharp contrast to restricted uptake of these elements by the iris,
hollyhock, and *Bupleurum*. The copper, lead, and zinc contents of
plant ash for species growing over the ore body was 1.5–4%, with a
simultaneous increase of *antimony*, gallium, and silver. Plant indicators
of ore bodies are: *Gypsophila*, *Astragalus*, *Euphrasia*, *Rumex acetosa*,
and *Centaurea montana*. There is a sharp decrease in numbers of
species growing over ore bodies, with one or two being predominant.
The effect of ore bodies on plant morphology is discussed and illustrated. The results of soil sampling, biogeochemistry, and geobotany
are compared and interpreted.

20-2   **ARSENIC**
Concentration in igneous rocks: 2 ppm
Concentration in soils: 5 ppm
Concentration in plant ash: 4 ppm
Toxicity to plants: moderate
Biogeochemical references: Berbenni (1959) (CA 53, 18354a), Minguzzi
   and Naldoni (1950) (CA 46, 2634b); Talipov et al. (1968) (CA 71,
   5374a); Warren and Delavault (1959) (CA 56, 3151a); Warren,
   Delavault, and Barakso (1964) (CA 62, 2624d)

ABSTRACT   Talipov et al. (1968). "Possible use of arsenic in biogeochemical prospecting for gold ore deposits." The *arsenic* of plant ash
from Southern Tamdytau is much higher than its clarke value for plants
of the earth's crust. A direct correlation was observed between the
arsenic and gold contents in plants and soils respectively. The *arsenic*
content of vegetation can be used during prospecting for gold. The
content of the former, was 10–15 times the normal background level in
nonmineralized areas.

20-3   **BARIUM**
Concentration in igneous rocks: 640 ppm
Concentration in soils: 500 ppm
Concentration in plant ash: 280 ppm

Toxicity to plants: moderate
Biogeochemical references: Bowen and Dymond (1955) (CA 50, 3693i); Gamble (1964) (CA 62, 5834b); Glazovskaya (1964) (CA 62, 4562h); Robinson et al. (1950) (CA 47, 9533i).

ABSTRACT   Glazovskaya (1964). "Biological cycle of elements in various landscape zones of the Urals." Trace elements in soils and plants from the Urals were analyzed by emission spectrography. The following enrichment coefficients (Kb) were obtained: *barium*, lead, strontium, $Kb = 100$; manganese and zinc, $Kb = 10$; molybdenum and tin, $Kb = 1$; chromium, nickel, and titanium, $Kb = 0.1$; arsenic and vanadium, $Kb = 0.01$. Enrichment coefficients decreased in a north-south direction probably because of an increase in pH. Biogenic elements (including *barium*) were enriched 3–10 times in the A horizons of soils.

20-4   **BERYLLIUM**
Concentration in igneous rocks: 4.2 ppm
Concentration in soils: 6 ppm
Concentration in plant ash: 0.7 ppm
Toxicity to plants: severe
Biogeochemical references: Cohen et al. (1967) (CA 68, 5011t); Kuzin (1959); Nikonova (1967) (CA 69, 33516a); Romney and Childress (1965) (CA 63, 18981e); Zalashkova et al. (1958) (CA 53, 5991c)

ABSTRACT   Nikonova (1967). "Storage of beryllium, molybdenum, zirconium, yttrium and other rare earth elements in vegetation in the southwestern Urals." Accumulation of the above elements was investigated by the biogeochemical method in order to locate ore deposits. *Beryllium* occurs in the wood of birch, larch, aspen, and pine and in *Vicia sylvatica*, *Aconitum excelsum*, and *Calamagrostis arundinacea* at a level of up to 10 ppm. Molybdenum is accumulated to the level of 10–30 ppm. *C. arundinacea*, an indicator for molybdenum, contains up to 300 ppm of this element. Zirconium is present mainly in larch and birch wood, *C. arundinacea* and *Dactylis glomerata* (10–100 ppm). Yttrium occurs in birch, larch, and leguminous plants (10–30 ppm). Titanium occurs in all tree and plant specimens (av. 300 ppm). Niobium was found only in *C. arundinacea* (up to 60 ppm). Lithium is concentrated up to 1000 ppm in *Filipendula ulmaria* and in pine.

20-5   **BISMUTH**
Concentration in igneous rocks: 0.2 ppm
Concentration in soils: 0.8 ppm
Concentration in plant ash: 1.2 ppm
Toxicity to plants: moderate
Biogeochemical reference: Hanna and Grant (1962) (CA 59, 5492c)

20-6   **BORON**
Concentration in igneous rocks: 13 ppm
Concentration in soils: 10 ppm
Concentration in plant ash: 700 ppm
Toxicity to plants: slight
Geobotanical references: Burenkov and Kuzina (1968); Buyalov and
    Shvyryayeva (1961); Kantor (1959) (CA 57, 4367e); Koval'sky,
    Ananichev, and Shakhova (1965) (CA 64, 10148c); Serdyuchenko
    and Pavlov (1967) (CA 70, 30684p)
Biogeochemical references: Ananichev (1960) (CA 55, 15641e); Buren-
    kov and Kuzina (1968); Kantor (1959) (CA 57, 4367e); Koval'sky,
    Ananichev, and Shakhova (1965) (CA 64, 10148c); Poluzerov (1965)
    (CA 63, 10607h); Robinson and Edgington (1942) (CA 36, 3826⁹);
    Selivanov (1946); Serdyuchenko and Pavlov (1967) (CA 70, 30684p);
    Shakhova (1960) (CA 55, 15558a)

ABSTRACT   Buyalov and Shvyryayeva (1961). "Geobotanical method
in prospecting for salts of boron." The successful use of geobotanical
indicators of *boron* is described. At very high concentrations of this
element, plant life is almost absent and is confined to *Salsola nitraria*
and *Limonium suffruticosum*. Where the magnesium content of
ascharites counteracts the effect of *boron*, *Eurotia ceratoides* associations indicate sulfate-carbonate deposits. Where the *boron* content of
the soil is of the order of several thousand ppm, plants are oppressed,
deformed, and subject to disease. A low *boron* content stimulates
plant growth.

20-7   **CADMIUM**
Concentration in igneous rocks: 0.13 ppm
Concentration in soils: 0.5 ppm
Concentration in plant ash: 0.1 ppm
Toxicity to plants: moderate
Biogeochemical reference: Hanna and Grant (1962) (CA 59, 5492c)

20-8   **CHROMIUM**
Concentration in igneous rocks: 117 ppm
Concentration in soils: 200 ppm
Concentration in plant ash: 9 ppm
Toxicity to plants: severe
Biogeochemical references: Glazovskaya (1964) (CA 62, 4562h);
    Karamata (1967) (CA 68, 97618x); Leutwein and Pfeiffer (1954) (CA
    49, 2960d); Lyon et al. (1968) (CA 69, 95506f); Paribok and
    Alekseeva-Popova (1966) (CA 64, 20564h); Porutsky et al. (1962)
    (CA 58, 2652g); Tyuremnov et al. (1968) (CA 69, 8901q); Vorob'ev
    and Gudoshnikov (1967) (CA 67, 10763y)

ABSTRACT   Lyon et al. (1968). "Trace elements in a New Zealand serpentine flora." Plants of a New Zealand serpentine flora together with their substrates were analyzed for *chromium*, cobalt, copper, and nickel. *Cassinia vauvilliersii* showed a highly significant correlation between plant ash and soil concentrations for all elements except copper. *Hebe odora* and *Leptospermum scoparium* showed similar but less pronounced correlations. *Pimelea suteri* was a strong accumulator of cobalt and nickel and *L. scoparium* for *chromium*. Calcium/magnesium ratios were higher for species growing over serpentine than for those over other soils. The above species, particularly *C. vauvilliersii*, should be useful for biogeochemical prospecting.

20-9   **COBALT**
Concentration in igneous rocks: 18 ppm
Concentration in soils: 10 ppm
Concentration in plant ash: 9 ppm
Toxicity to plants: severe
Geobotanical references: Duvigneaud (1958, 1959); Storozheva (1954)
   (CA 49, 11110e)
Biogeochemical references: Cabrol et al. (1961) (CA 60, 7706i);
   Duvigneaud (1958, 1959); Lyon et al. (1968) (CA 69, 95506f);
   Malyuga (1950) (CA 45, 4757c), (1951) (CA 45, 3972g); Malyuga
   and Petrunina (1961) (CA 56, 5667e); Paribok and Alekseeva-
   Popova (1966) (CA 64, 20564h); Poskotin (1963) (CA 62, 342g);
   Tooms and Jay (1964) (CA 61, 13054e); Tyuremnov et al. (1968)
   (CA 69, 8901q); Vekilova et al. (1963) (CA 61, 9597e); Vorob'ev
   and Gudoshnikov (1967) (CA 67, 10763y); Warren and Delavault
   (1957) (CA 52, 13554b)

ABSTRACT   Duvigneaud (1959). "Plant cobaltophytes in Upper Katanga." There is a plant association in Upper Katanga linked with copper-rich soils containing abnormal amounts of *cobalt* (up to 2000 ppm). Most of the vegetation is cuprophile but *Crotalaria cobalticola* and *Silene cobalticola* appear to be associated with *cobalt* only and contain up to 17,700 ppm in the ash. The high *cobalt* content of the soils also has the effect of limiting numbers of species over mineralized areas since many cuprophile plants of the area will not tolerate abnormal amounts of *cobalt*.

20-10   **COPPER**
Concentration in igneous rocks: 70 ppm
Concentration in soils: 20 ppm
Concentration in plant ash: 180 ppm
Toxicity to plants: severe

Geobotanical references: Bailey (1899); Chou-Chin-Han (1960); Cole et al. (1968); Duvigneaud (1958, 1959); Hartman (1969); Henwood (1857); Horizon (1959); Lovering et al. (1950) (CA 45, 78b); Malyuga et al. (1959) (CA 54, 7445i); Nesvetaylova (1961); Noguchi (1956); Persson (1948, 1956); Robyns (1932); Schatz (1955); Shacklette (1961), (1965a) (CA 64, 5450e), (1965b), (1967) (CA 68, 14783y); Skertchly (1897); Tsung-Shan (1957); Vogt (1942a) (CA 41, 667h)

Biogeochemical references: Basitova et al. (1964) (CA 63, 9672d); Brooks and Lyon (1966) (CA 66, 12935j); Chiu and Wang (1964); Cole et al. (1968); Duvigneaud (1958, 1959); Flerova and Flerov (1964) (CA 62, 14351e); Fortescue and Hornbrook (1969) (CA 71, 93426f); Harbaugh (1950) (CA 45, 716e); Hornbrook (1969) (CA 71, 93458j); Khamrabaev and Talipov (1960) (CA 55, 8196c); Lovering et al. (1950) (CA 45, 78b); Malyuga (1950) (CA 45, 4757c), (1951) (CA 45, 3972g), (1954) (CA 49, 10809f), (1958b) (CA 54, 25456a), (1959) (CA 53, 12101i); Malyuga and Makarova (1955) (CA 53, 19251d); Malyuga et al. (1960) (CA 55, 13191h); Malyuga and Petrunina (1961) (CA 56, 5667e); Marmo (1953) (CA 48, 11990g); Nicolas and Brooks (1969); Nicolls et al. (1965) (CA 63, 16048d); Poskotin (1963) (CA 62, 342g); Poskotin and Lyubimova (1963) (CA 64, 12380f); Riddell (1952); Salmi (1956) (CA 51, 7256f); Skertchly (1897); Starikov et al. (1964) (CA 61, 15847d); Talipov (1964a) (CA 61, 4103b), (1964b) (CA 62, 6268h), (1965) (CA 64, 12379g), (1966) (CA 68, 71346g); Timperley et al. (1970a, 1970b); Vogt and Braadlie (1942) (CA 41, 668b); Vogt and Bugge (1943) (CA 39, 3763[4]); Vogt et al. (1943) (CA 39, 3763[5]); Warren and Delavault (1948) (CA 43, 3321c), (1949) (CA 43, 3321h), (1955b), (1955c) (CA 50, 7681h), (1960) (CA 55, 16305h); Warren, Delavault, and Cross (1959); Warren, Delavault, and Irish (1949) (CA 44, 6787d); (1951) (CA 46, 3914g); (1952a) (CA 46, 6555f); Warren and Howatson (1947) (CA 42, 1533c); White (1950) (CA 44, 10614c); Worthington (1955) (CA 50, 1536g)

ABSTRACT   Nicolls et al. (1965). "Geobotany and biogeochemistry in mineral exploration in the Dugald River area, Cloncurry district, Australia." The vegetation assemblages of the Dugald River area correspond clearly with outcrops of the lithologically distinct geologic units. *Scaevola densevestita*, *Ptilotus obovatus*, and *Cassia desolata* grow over *copper* anomalies and over calcareous rocks. *Tephrosia* sp. seems to be a good indicator of soil *copper* anomalies greater than 2000 ppm. It also grows over lead and zinc anomalies greater than 1000 ppm. *Polycarpaea glabra* grows over *copper* anomalies greater than 2000 ppm and also over strong lead-zinc anomalies. The most efficient procedure would involve a combination of geobotanical and geochemical prospecting techniques.

20-11   **GALLIUM**

Concentration in igneous rocks: 20 ppm
Concentration in soils: 20 ppm
Concentration in plant ash: 1 ppm
Toxicity to plants: slight
Biogeochemical references: Dobrovol'sky (1963) (CA 60, 1064f); Flerova and Flerov (1964) (CA 62, 14351f)

ABSTRACT   See 20–1.

20-12   **GERMANIUM**

Concentration in igneous rocks: 2 ppm
Concentration in soils: 5 ppm
Concentration in plant ash: 5 ppm
Toxicity to plants: slight
Biogeochemical references: Geilmann and Brünger (1935); Paul (1953); Tchakirian (1942) (CA 38, 2580[9])

ABSTRACT   Paul (1953). "Germanium in contemporary vegetation." *Germanium* is found in plant ash to the extent of several hundred ppm in some cases. The occurrence of this element in plants is linked to drainage. *Germanium* is very readily leached from freshly harvested vegetation which should be analyzed immediately.

20-13   **GOLD**

Concentration in igneous rocks: 0.001 ppm
Concentration in soils: 0.002 ppm
Concentration in plant ash: 0.005 ppm
Toxicity to plants: slight
Biogeochemical references: Aferov et al. (1968) (CA 70, 30889j), Aripova and Talipov (1966) (CA 66, 48364m), Babička (1943) (CA 39, 2770[7]), Cannon et al. (1968) (CA 69, 103881s); Lobanov, Khatamov, and Khamidova (1967) (CA 69, 631f); Lobanov, Khatamov, and Kist (1967) (CA 68, 119123m); Lobanov et al. (1966) (CA 69, 29220g); Nemec et al. (1936); Razin and Roshkov (1963) (CA 59, 9691b); Safronov et al. (1958) (CA 55, 1307b); Talipov et al. (1968) (CA 71, 5374a); Warren and Delavault (1950a) (CA 44, 4086a)

ABSTRACT   Lobanov et al. (1966). "Efficiency of neutron activation logging during biogeochemical prospecting for gold deposits in Central Kazakhstan." Interference effects in the determination of *gold* in plant ash and soils by neutron activation are controlled by irradiating samples in cadmium containers. Over 300 samples of plants and soils were analyzed. A maximum *gold* content of 36 ppm was found in the ash of

Lagochilus intermedius. *Gold* accumulated mainly in grasses and bushes, predominantly in leaves and small branches. Relatively high *gold* contents were observed in the leaves of *Salsola rigida*, *Haplophyllum robustum*, and *Girgensohnia*, and in upper branches of *S. arbuscula*, *S. carinata*, and *Cryptodiscus didymus*. There was no direct correlation between the *gold* contents of plants and soils. The biogeochemical method is the most efficient for prospecting in Central Asia where the soil sampling method is not applicable because of the low *gold* content of the upper soil horizons.

20-14   IRON
Concentration in igneous rocks: 4.65%
Concentration in soils: 1–5%
Concentration in plant ash: 6700 ppm
Toxicity to plants: slight
Geobotanical references: Buck (1951); Le Jolis (1860); Lidgey (1897)
Biogeochemical references: Mizuno (1967b) (CA 68, 38559f); Tkalich (1953); Warren, Delavault, and Irish (1952b) (CA 48, 11259a)

ABSTRACT   Warren, Delavault, and Irish (1952b). "Preliminary studies on the biogeochemistry of iron and manganese." Studies of *iron* and manganese contents of leaves and twigs were made in order to find variations which might indicate mineralization. *Iron* and manganese variations were studied in different species, different organs of the same tree, under different climatic conditions, over different geologic formations, and in pine twigs from the area of the Sullivan Mine. It is suggested that *iron* and manganese variations might be most useful in areas where gold and silver are accompanied by high contents of *iron* and manganese but not copper or zinc.

20-15   LANTHANIDES (RARE EARTHS) AND YTTRIUM
Concentration (lanthanum) in igneous rocks: 45 ppm
Concentration (lanthanum) in soils: 40 ppm
Concentration (total rare earths) in plant ash: 16 ppm
Toxicity to plants: slight
Biogeochemical references: Mitskevich (1962) (CA 60, 13026g); Nikonova (1967) (CA 69, 33516a); Robinson (1943) (CA 37, 6791[8]), (1951) (CA 46, 6678f); Robinson et al. (1958) (CA 53, 148c); Robinson and Edgington (1945) (CA 40, 419[2]); Robinson et al. (1938) (CA 32, 5445[4]); Talipov et al. (1967) (CA 68, 104194m)

ABSTRACT   Robinson et al. (1958). "Biogeochemistry of the rare earth elements with particular reference to hickory leaves." Hickory leaves concentrate *rare earths* from soils. Average percentages of total *rare earths* accumulated were: yttrium 36, lanthanum 16, cerium 14, praseodymium 2, neodymium 20, samarium 1, europium 0.7, gadolinium 3, terbium 0.6, dysprosium 3, holmium 0.7, erbium 2, thulium 0.2, ytterbium 1, and lutecium 0.2. Cerium showed large variations and had abnormally low concentrations compared with other *rare earths* probably because it is believed to exist in soils as Ce^{4+} which is less available to plants than other trivalent *lanthanides.*

20-16   LEAD

Concentration in igneous rocks: 16 ppm

Concentration in soils: 10 ppm

Concentration in plant ash: 70 ppm

Toxicity to plants: severe

Geobotanical references: Cannon (1960b) (CA 54, 22186c); Cole et al. (1968)

Biogeochemical references: Basitova et al. (1964) (CA 63, 9672d); Brown and Meyer (1960) (CA 57, 3097f); Cannon and Bowles (1962) (CA 58, 870f); Chapman and Shacklette (1960) (CA 55, 13191c); Cole et al. (1968); Flerova and Flerov (1964) (CA 62, 14351e); Fortescue and Hornbrook (1969) (CA 71, 93426f); Harbaugh (1950) (CA 716e); Hornbrook (1969) (CA 71, 93458j); Keith (1968); Lyubofeev et al. (1962) (CA 60, 7805e); Makarova (1960) (CA 55, 11199h); Malyuga and Makarova (1955) (CA 53, 19251d); Malyuga et al. (1960) (CA 55, 13191h); Nicolas and Brooks (1969); Nicolls et al. (1965) (CA 63, 16048d); Parfent'eva (1955); Poskotin (1963) (CA 62, 342g); Poskotin and Lyubimova (1963) (CA 64, 12380f); Salmi (1956) (CA 51, 7256f); Shacklette (1960) (CA 55, 13736h); Starikov et al. (1964) (CA 61, 15847d); Talipov (1964a) (CA 61, 4103b), (1964b) (CA 62, 6268h), (1965) (CA 64, 12379g); Timperley et al. (1970b); Warren (1966); Warren and Delavault (1960) (CA 55, 16305h); Webb and Millman (1951) (CA 45, 9428e)

ABSTRACT   Nicolas and Brooks (1969). "Biogeochemical prospecting for zinc and lead in the Te Aroha region of New Zealand." The uptake of *lead* and zinc by the vegetation of the Tui Base Mine area near Te Aroha was investigated. Statistical analysis of chemical data indicated that the *lead* content of twigs of *Beilschmiedia tawa* and the zinc concentration in leaves of *Schefflera digitata* correlated well with the amounts of these elements in the soil.

20-17   LITHIUM

Concentration in igneous rocks: 50 ppm

Concentration in soils: 30 ppm

Concentration in plant ash: 2 ppm

Toxicity to plants: slight

Biogeochemical references: Borovik-Romanova (1965) (CA 64, 7044e); Ezdakova (1964) (CA 62, 9455g), (1966) (CA 65, 7623g); Nikonova (1967) (CA 69, 33516a)

ABSTRACT   Borovik-Romanova (1965). "Content of lithium in soils." The content of *lithium* was determined in 38 samples of plants from 17 families and in the soils in which they grew. The *lithium* content of vegetation depended on the species and content of the element in the soil. In some plants, the enrichment coefficient was 0.01–0.04, while in others the value was 0.10–0.17 or greater. The following plants are in the latter group: *Thalictrum minus*, *T. flavum*, *T. simplex*, *T. contortum*, *T. angustifolium*, *Lycium barbarum*, *Solanum depilatum*, *Nicotinia* sp., *Datura stramonium*, *Atropa belladonna*, *Cirsium echinus*, *Adonis* sp., *Pulsatilla albana*, *Ranunculus repens*, *R. polyanthemos*, *R. illyricus*, *Physalis alkekengi*, *Scopolia carniolica*, and *Myosotis alpestris*. The *lithium* content of these plants also depended on the *lithium* content of the soil (20–30 ppm). Enrichment coefficients remained relatively constant and there was a good plant response to variations in the content of the element in the soil.

20-18   **MANGANESE**

Concentration in igneous rocks: 1000 ppm

Concentration in soils: 850 ppm

Concentration in plant ash: 4800 ppm

Toxicity to plants: moderate

Geobotanical reference: Uspensky (1915)

Biogeochemical references: Bloss and Steiner (1960) (CA 54, 19332e); Bradis et al. (1963) (CA 62, 820h); Chamberlain and Searle (1963) (CA 60, 9853e); Chartko (1967) (CA 68, 104175f); Gerloff et al. (1964) (CA 61, 12332d); Glazovskaya (1964) (CA 62, 4562h); Kashlev et al. (1966) (CA 67, 8683d); Kovacs (1966) (CA 67, 63336t); Levanidov and Khilyukova (1953); Liwski (1961) (CA 57, 5027b); Malyuga (1947) (CA 43, 3549e); Mehta et al. (1964) (CA 64, 7311f); Mizuno (1967b) (CA 68, 38559f); Muzaleva and Pershina (1965) (CA 65, 19254a); Paribok and Alekseeva-Popova (1966) (CA 64, 20564h); Podkorytov (1967) (CA 69, 103827d); Poluzerov (1965) (CA 63, 10607h); Shnyukov et al. (1963) (CA 61, 463d); Tolgyesi (1962) (CA 58, 3687a); Tyuremnov et al. (1968) (CA 69, 8901q); Warren, Delavault, and Irish (1952b) (CA 48, 11259a)

ABSTRACT   Bloss and Steiner (1960). "Biogeochemical prospecting for manganese in Northeast Tennessee." Chestnut oaks were analyzed for *manganese*, iron, barium, and nickel in prospecting for *manganese*. The mineralization appeared to be reflected in the chemical composition of the trees of the area, especially in the one area which was later developed into an economic ore body. Nickel appeared to be a reliable indicator for *manganese*. Above-average amounts of nickel were detected in vegetation above the ore body, though the soil contained normal amounts of this element.

20-19   **MERCURY**
Concentration in igneous rocks: 0.06 ppm
Concentration in soils: 0.01 ppm
Concentration in plant ash: 0.01 ppm
Toxicity to plants: severe
Biogeochemical references: Karasik (1962) (CA 60, 11762b); Nikiforov and Fedorchuk (1959) (CA 55, 12173i); Warren, Delavault, and Barakso (1966) (CA 66, 4830k)

ABSTRACT   Warren, Delavault, and Barakso (1966). "Some observations on the geochemistry of mercury as applied to prospecting." Around *mercury* deposits there are usually 1–10 ppm of the element in soils. Different organs and different species of plants vary widely in their ability to concentrate *mercury*. In medium and low ranges, plant sampling offers some advantages over soil sampling. If the calculated ash content is greater than 10 ppm *mercury*, there is probable mineralization in the area.

20-20   **MOLYBDENUM**
Concentration in igneous rocks: 1.7 ppm
Concentration in soils: 2.5 ppm
Concentration in plant ash: 13 ppm
Toxicity to plants: moderate
Geobotanical references: Koval'sky and Petrunina (1964) (CA 62, 8343d); Robinson and Edgington (1948) (CA 43, 8883i)
Biogeochemical references: Aripova and Talipov (1966) (CA 66, 48364m); Bradis et al. (1963) (CA 62, 820h); Brooks and Lyon (1966) (CA 66, 12935j); Cannon (1960c) (CA 55, 14179a); Glazovskaya (1964) (CA 62, 4562h); Il'in (1966) (CA 67, 2423x); Kabata-Pendias and Bolibrzuch (1964) (CA 61, 15294e); Kazitsyn and Aleksandrov (1960) (CA 56, 715b); Koval'sky and Yarovaya (1966) (CA 66, 1882e); Liwski (1961) (CA 57, 5027b); Lyon and Brooks (1969); Malyuga (1958a) (CA 52, 18108b), (1958b) (CA 54, 25456a), (1959) (CA 53, 12102i), (1960) (CA 55, 15797g), Mizuno (1967a) (CA 68, 38561a), Muzaleva and Pershina (1965) (CA 65, 19254a), Nikonova (1967) (CA 69, 33516a); Poluzerov (1965) (CA 63, 10607h); Talipov (1964b) (CA 61, 4103c), (1965) (CA 64, 12379g); Timperley et al. (1970b); Warren and Delavault (1965); Warren,

Delavault, and Routley (1953) (CA 48, 5748f); Yakovleva (1964) (CA 63, 2767b); Yarovaya (1960) (CA 55, 15558a)

ABSTRACT   Malyuga (1959). "Application of the biogeochemical method in prospecting for copper-molybdenum ores." The *molybdenum* content of plants and soils correlates with the topography of the Kadzharan sulfide deposits. The copper and *molybdenum* contents of plants and soils were a hundred times greater than background over the ore deposits. A direct plant-soil correlation was established between copper and *molybdenum*. Maps were obtained of isoconcentrations of *molybdenum* in plants and soils and allowed the outline of the ore body to be delineated.

20-21   NICKEL

Concentration in igneous rocks: 100 ppm
Concentration in soils: 40 ppm
Concentration in plant ash: 65 ppm
Toxicity to plants: severe
Geobotanical references: Minguzzi and Vergnano (1948) (CA 44, 9003h), Storozheva (1954) (CA 49, 11110e)
Biogeochemical references: Aleskovsky et al. (1959) (CA 53, 18775h); Dobrovol'sky (1963) (CA 60, 1064f); Fortescue and Hornbrook (1969) (CA 71, 93426f); Glazovskaya (1964) (CA 62, 4562h); Grigoryan (1966) (CA 66, 57911q); Hornbrook (1969) (CA 71, 93458j); Leutwein and Pfeiffer (1954) (CA 49, 2960d); Lyon et al. (1970); Lyon et al. (1968) (CA 69, 95506f); Malyuga (1939) (CA 34, 4021[8]), (1954) (CA 49, 10809f); Malyuga and Makarova (1955) (CA 53, 19251d); Miller (1961) (CA 56, 9748i); Minguzzi and Vergnano (1948) (CA 44, 9003h); Mizuno (1967a) (CA 68, 38561a); Paribok and Alekseeva-Popova (1966) (CA 64, 20564h); Porutsky et al. (1962) (CA 58, 2652g); Shacklette (1962a) (CA 56, 15831d); Sokolova and Yatsyuk (1965) (CA 65, 20511g); Timperley et al. (1970a, b); Tyuremnov et al. (1968) (CA 69, 8901q); Vekilova et al. (1963) (CA 61, 9597e); Vorob'ev and Gudoshnikov (1967) (CA 67, 10763y); Warren and Delavault (1955a) (CA 49, 6382f)

ABSTRACT   Leutwein and Pfeiffer (1954). "The possible usefulness of geochemical prospecting methods for hydrous nickel silicate ores." Analysis for a number of elements including *nickel*, copper, and zinc was carried out on plants, soils, and humus from a nickeliferous area in Germany. Plants analyzed were mainly *Betula verrucosa*, *Picea abies*, and *Pinus sylvestris*. The results indicated that the biogeochemical method could be used as a method of prospecting since *nickel* contents of over 300 ppm in soils, over 500 ppm in the ash of pine bark, over 600 ppm in the ash of birch leaves, and over 1000 ppm in spruce bark, were observed only over rocks enriched in *nickel*. Cobalt and chromium could also be used as indicators for *nickel*.

20-22   **NIOBIUM**

Concentration in igneous rocks: 20 ppm
Concentration in soils: 15 ppm
Concentration in plant ash: 0.3 ppm
Toxicity to plants: not known but probably slight
Biogeochemical references: Mitskevich (1962) (CA 60, 13026g); Nikonova (1967) (CA 69, 33516a); Tyutina et al. (1959) (CA 54, 7459i); Vakhromeev (1962) (CA 59, 11121a)

ABSTRACT   Tyutina et al. (1959).   "Practice of biogeochemical prospecting and procedure for determination of niobium in plants."   Biogeochemical prospecting was carried out in the Central Timan area, Komi, ASSR, where rare metal mineralization was associated with an elevated *niobium* content of the soil.   This element was determined in 10–40 specimens for each species studied.   Elevated *niobium* contents were observed in *Rubus arcticus*, *Vaccinium myrtillus*, *Chamaenerion angustifolium*, *Betula pubescens*, and *B. verrucosa*.   Low *niobium* contents were found in *Sphagnum*, *Salix* sp., *Picea obovata*, *Carex vesicaria*, and *Ribes nigrum*.   More than 50% of the species contained 0–0.60 ppm *niobium* in leaves (dry weight).   The absorption of *niobium* in the organs of each species was studied and was higher in leaves and branches than stems.   In the roots of *Chamaenerion angustifolium*, the content was 2–3 times greater than in the leaves.

20-23   **RADIUM**

Concentration in igneous rocks: 9×10^{-7} ppm
Concentration in soils: 8×10^{-7} ppm
Concentration in plant ash: 2×10^{-8} ppm
Toxicity to plants: severe (due to radioactivity)
Geobotanical reference: Sokolova and Khramova (1961) (CA 57, 8229h)
Biogeochemical references: Baranov and Kunasheva (1954) (CA 49, 11110f); Grodzinsky (1959) (CA 54, 16556d); Grodzinsky and Golubkova (1964) (CA 63, 16750a); Gvozdanovic et al. (1967) (CA 69, 21188g); Kovalevsky (1962a) (CA 57, 14182i), (1962b) (CA 58, 13079c), (1964) (CA 63, 2794b), (1965) (CA 63, 13706h); Kunasheva (1944) (CA 40, 5332[7]); Makarov (1965) (CA 64, 456c); Sokolova and Khramova (1961) (CA 57, 8229h); Verkhovskaya et al. (1967) (CA 68, 36448v)

ABSTRACT   Makarov (1965). "Biogeochemical prospecting for uranium in the USSR." The alpha activity of plant ash was used to prospect for uranium in the Soviet Union. The radioactivity is derived from *radium* halos. The equilibrium between *radium*-226 and radon, and the presence of *radium* derived from thorium minerals should be taken into account. Measurement of the radioactivity of vegetation without ashing was also used as an exploratory tool. Various plant species and organs accumulated *radium* in a different manner. The reliability of the method was checked by gamma scintillometry, drilling, and analysis for uranium. For mapping in the range 1:10,000 to 1:100,000, sample spacing should be 1–20 and 25–50 meters. For more detailed prospecting, sample intervals should be 5–10 meters. The method can be used during the cold season when other techniques are impracticable.

20-24   RHENIUM
Concentration in igneous rocks: 0.005 ppm
Concentration in soils: 0.005 ppm
Concentration in plant ash: 0.005 ppm
Toxicity to plants: slight
Biogeochemical reference: Cannon (1964) (CA 62, 6268a); Myers and
    Hamilton (1960), (1961) (CA 56, 10586i)

ABSTRACT   Myers and Hamilton (1960). "Rhenium in plant samples from the Colorado Plateau." *Rhenium* was analyzed in species of *Astragalus* and *Eriogonum* growing in uraniferous ground in the Gypsum Valley area of Colorado. The concentrations of this element ranged from 50 to 500 ppm in the ash. Samples of other species growing over a schroekingerite deposit in the Yellow Cat area, Utah, were also analyzed for *rhenium*. The highest value of 300 ppm was found in the ash of *Atriplex confertifolia*. A specimen of *Oenothera caespitosa* from Grant County, New Mexico, gave a value of 150 ppm *rhenium*. The data indicate that this element is concentrated preferentially in leaves rather than stems or roots.

20-25   SELENIUM
Concentration in igneous rocks: 0.01 ppm
Concentration in soils: 0.5 ppm
Concentration in plant ash: 1 ppm
Toxicity to plants: moderate
Geobotanical references: Cannon (1952) (CA 47, 10638f), (1953) (CA
    47, 6315a), (1957) (CA 52, 2677d), (1959) (CA 56, 3150i), (1960a)

(CA 55, 12177f), (1960c) (CA 55, 14179a), (1964) (CA 62, 6268a); Cannon and Kleinhampl (1956) (CA 50, 16584g); Cannon and Starrett (1956) (CA 50, 9247b); Froelich and Kleinhampl (1960) (CA 55, 2372b); Kleinhampl (1962); Kleinhampl and Koteff (1960) (CA 53 (13905b); McCray and Hurwood (1963) (CA 61, 12539h); Miller and Byers (1937) (CA 31, 7164[8]); Trelease and Beath (1949) (CA 44, 3348a)

Biogeochemical references: Johnson et al. (1966) (CA 68, 47023d); Koval'sky and Ermakov (1967) (CA 66, 57928a); Leutwein and Starke (1957) (CA 51, 17646f); McCray and Hurwood (1963) (CA 61, 12539h); Miller and Byers (1937) (CA 31, 7164[8]); Trelease and Beath (1949) (CA 44, 3348g)

ABSTRACT Cannon (1957). "Description of indicator plants and methods of botanical prospecting for uranium deposits on the Colorado Plateau." Mapping indicator plants was used in prospecting for uranium in arid parts of the Colorado Plateau. The distribution of indicator plants is controlled by the availability of chemical constituents of the ore such as *selenium*, sulfur, and calcium. Plants of the genus *Astragalus* are most useful in prospecting for uranium deposits of high *selenium* content. Fifty indicator plants are associated with carnotite and included 14 which are directly controlled by *selenium*.

20-26   SILVER

Concentration in igneous rocks: 0.2 ppm

Concentration in soils: 1 ppm

Concentration in plant ash: 1 ppm

Toxicity to plants: severe

Geobotanical references: Bailey (1899); Henwood (1857)

Biogeochemical references: Bradis et al. (1963) (CA 62, 820h); Cannon et al. (1968) (CA 69, 103881s); Dobrovol'sky (1963) (CA 60, 1064f); Flerova and Flerov (1964) (CA 62, 14351f); Polikarpochkina et al. (1965) (CA 65, 1976b); Poluzerov (1965) (CA 63, 10607h); Warren and Delavault (1950a) (CA 44, 4086a); Webb and Millman (1951) (CA 45, 9428c)

ABSTRACT Warren and Delavault (1950a). "Gold and silver content of some trees and horsetails (*Equisetum* spp.) in British Columbia." Gold and *silver* were determined by fire assay in leaves, twigs, and needles collected from the Clinton Mining district. Up to 0.07 ppm gold was found in *Equisetum* spp. and up to 0.02 ppm in common trees growing in auriferous areas. In background areas, trees and lesser plants growing over areas containing no known *silver* mineralization contained 0.1–1.4 ppm of this element.

20-27   **STRONTIUM**
Concentration in igneous rocks: 350 ppm
Concentration in soils: 300 ppm
Concentration in plant ash: 30 ppm
Toxicity to plants: slight
Geobotanical reference: Koval'sky and Zasorina (1965) (CA 63, 15488c)
Biogeochemical references: Basitova and Zasorina (1965) (CA 64, 6342c); Bowen and Dymond (1955) (CA 50, 3693i); Koval'sky, Basitova, and Zasorina (1965) (CA 63, 18977g); Koval'sky et al. (1968) (CA 70, 98806k); Koval'sky and Zasorina (1965) (CA 63, 15488c)

ABSTRACT   Koval'sky and Zasorina (1965). "Biogeochemical aspect of strontium." *Strontium* and calcium were determined in plants and soils of Southern Tadzhikstan. Some species such as *Echium italicum*, *Alhagi kirghisorum*, *Ampelopsis vitifolia*, and *Glycyrrhiza glabra*, accumulate *strontium* actively and contain 0.21–2.5% (dry weight). All *Gramineae* contain 26–410 ppm (dry weight) of this element. No morphological changes were detected in plants growing over strontium-rich soils, but some anatomical and physiological changes were observed in *Alhagi kirghisorum*, *Salvia officinalis*, and *Capparis spinosa*. Some plant species revealed some degree of adaptation to the increased content of *strontium* in the soil.

20-28   **THORIUM**
Concentration in igneous rocks: 13 ppm
Concentration in soils: 13 ppm
Concentration in plant ash: 20 ppm
Toxicity to plants: slight
Biogeochemical references: Baranov and Kunasheva (1954) (CA 49, 11100f); Grodzinsky (1959) (CA 54, 16556d); Gvozdanovic et al. (1967) (CA 69, 21188g); Lobanov and Khatamov (1968) (CA 70, 84942g); Verkhovskaya et al. (1967) (CA 68, 36448v)

ABSTRACT   Verkhovskaya et al. (1967). "The migration of natural radioactive elements under natural conditions and their distribution according to biotic and abiotic environmental conditions." The content and distribution of radium, uranium, and *thorium* in soils, vegetation, and animals is given for zones of northern taiga. Plants remove soluble forms of these three elements but their content does not reflect their concentrations in the soil. An acropetal gradient is found for the concentrations of the three elements in plants and the amounts increase with the age of the plant specimen.

20-29   **TIN**
Concentration in igneous rocks: 32 ppm
Concentration in soils: 10 ppm
Concentration in plant ash: 1 ppm
Toxicity to plants: severe
Biogeochemical references: Dobrovol'sky (1963) (CA 60, 1064f); Glazovskaya (1964) (CA 62, 4562h); Ivashov and Bardyuk (1967); Lindner and Sarosiek (1963) (CA 61, 6808c); Millman (1957) (CA 51, 17639b); Paribok and Alekseeva-Popova (1966) (CA 64, 20564h); Polikarpochkina et al. (1965) (CA 65, 1976b); Sarosiek and Klys (1962) (CA 60, 16451h); Zakirov et al. (1959) (CA 53, 13276i)

ABSTRACT   Sarosiek and Klys (1962). "The tin content of plants and soils of Sudety." The highest *tin* content was found in agricultural soil in Gierczyn, and the lowest (51 ppm) in the natural soils of Pobiedna. In the latter, the *tin* content was highest in the humus level. The following plant species accumulated *tin*: *Calluna vulgaris*, *Gnaphalium sylvaticum*, *Sempervivium soboliferum*, *Silene inflata*, *Tanacetum vulgare*, and *Quercus sessilis*. The average content of the *tin* in the ash of these plants was 46 ppm.

20-30   **TUNGSTEN**
Concentration in igneous rocks: 2 ppm
Concentration in soils: 1 ppm
Concentration in plant ash: 0.5 ppm
Toxicity to plants: moderate
Biogeochemical references: Dobrolyubsky and Vozdeistvy (1964) (CA 61, 8850e); Kabiashvili (1964); Kovalevsky (1966b) (CA 65, 5872d); Polikarpochkina et al. (1965) (CA 65, 1976b); Zakirov et al. (1959) (CA 53, 13276i)

ABSTRACT   Kovalevsky (1966b). "Biogeochemistry of tungsten in plants." *Tungsten* is concentrated by plants to a significant degree and is enriched in the older plant organs. The ash of the plant organs usually contains between a tenth to a half of the amount in the soil. The concentration of *tungsten* in the older organs of woody plants correlates with its content in the B and C horizons of the soil. There is no correlation with leaves. For this reason, tree trunks or twigs which are two or three years old should be sampled in biogeochemical prospecting. Grasses are also suitable for this purpose. If sampling patterns are close enough (5 to 10 m apart), the biogeochemical method may be used for discovering individual ore veins.

20-31   URANIUM
Concentration in igneous rocks: 2.6 ppm
Concentration in soils: 1 ppm
Concentration in plant ash: 0.6 ppm
Toxicity to plants: moderate
Geobotanical references: Cannon (1952) (CA 47, 10638f), (1957) (CA
    52, 2677d), (1960a) (CA 55, 12177f), (1964) (CA 62, 6268a);
    Shacklette (1962a) (1964).
Biogeochemical references: Anderson and Kurtz (1954), (1955), (1956)
    (CA 50, 13672c); Armands and Landergren (1960) (CA 54, 24170h);
    Botova et al. (1963) (CA 59, 1389d); Cannon (1952) (CA 47, 10638f),
    (1953) (CA 47, 6315a), (1957) (CA 52, 2677d), (1959) (CA 56,
    3150i), (1960a) (CA 55, 12177f), (1960c) (CA 55, 14179a), (1964)
    (CA 62, 6268a); Cannon and Kleinhampl (1956) (CA 50, 16584g);
    Cannon et al. (1968) (CA 69, 103881s); Cannon and Starrett (1956)
    (CA 50, 9247b); Debnam (1955); Froelich and Kleinhampl (1960)
    (CA 55, 2372b); Goldzstein (1957) (CA 52, 7961i); Grodzinsky
    (1959) (CA 54, 16556d); Grodzinsky and Golubkova (1964) (CA 63,
    16750a); Kleinhampl (1962); Kleinhampl and Koteff (1960) (CA 53,
    13905b); Konstantinov (1963) (CA 59, 2518a); Kovalevsky (1962a)
    (CA 57, 14182i), (1962b) (CA 58, 13079c), (1964) (CA 63, 2794b),
    (1965) (CA 63, 13706h), (1966a) (CA 67, 40828r); Koval'sky and
    Voronitskaya (1966)(CA 65, 20779e); Lecoq et al. (1958) (CA 53,
    4035a); Lobanov and Khatamov (1968) (CA 70, 84942g); Makarov
    (1965) (CA 64, 456c); Mamulea and Buracu (1967) (CA 68, 14790y);
    Moiseenko (1959) (CA 53, 9921c); Murakami et al. (1958) (CA 53,
    2959e); Ohashi and Murakami (1960) (CA 56, 12571f); Rackley et al.
    (1968) (CA 70, 108138k); Sokolova and Khramova (1961) (CA 57,
    8229h); Verkhovskaya et al. (1967); Vostokova (1957) (CA 52, 183c);
    Whitehead and Brooks (1969a) (CA 70, 108236r); (1969b), (1969c),
    (1971a), (1971b); Yakloleva (1964)(CA 63, 2767b)

ABSTRACT   Cannon (1964). "Geochemistry of rocks and related soils
and vegetation of the Yellow Cat area, Grand County, Utah." Sulfur,
selenium, arsenic, and molybdenum are concentrated along with
uranium and vanadium in the ores of the Yellow Cat area. A geochemical halo of these six elements envelopes each ore body. Two
botanical methods of prospecting were used in the area. One of these
involved the analysis of juniper needles and leaves of shrubs. The
uranium content of plant ash was 0.5 ppm for barren ground and 2 ppm
for mineralized areas. The other method involved the mapping of the
distribution of indicator plants. Two selenium accumulators, *Astragalus preussi* and *A. pattersoni*, proved to be especially good indicators
of mineralized ground. After completion of plant studies, 1268 holes
drilled in the area of plant mapping verified the presence of ore.

The selenium indicators grew on 81% of the ground mineralized at a depth of less than 32 ft and in 42% of that mineralized at a depth of 32–170 ft.

20-32   VANADIUM

Concentration in igneous rocks: 90 ppm
Concentration in soils: 100 ppm
Concentration in plant ash: 22 ppm
Toxicity to plants: moderate
Geobotanical reference: Cannon (1964) (CA 62, 6268a).
Biogeochemical references: Bertrand (1950); Cannon (1952) (CA 47, 10638f), (1960a)(CA 55, 12177f), (1960c)(CA 55, 14179a), (1963) (CA 61, 5926f), (1964)(CA 62, 6268a); Dobrovol'sky (1963)(CA 60, 1064f); Glazovskaya (1964)(CA 62, 4562h); Lindner and Sarosiek (1963) (CA 61, 6808c); Porutsky et al. (1962)(CA 58, 2652g); Waring-ton (1956)(CA 51, 8902c).

ABSTRACT   Cannon (1963).   "The biogeochemistry of vanadium." Small amounts of *vanadium* are stimulating to plants and large amounts are toxic. Ten to 20 ppm *vanadium* in nutrient solutions is commonly harmful to plants but larger amounts are tolerated by legumes that use *vanadium* in nitrogen fixation processes. *Allium* and some species of *Astragalus*, *Castilleja*, and *Chrysothamnus* are accumulators of *vanadium*. The *vanadium* content of aerial parts of plants rooted in calcic soils is low and that of plants rooted in seleniferous soils is very high. Plants growing in calcic soils are very tolerant of *vanadium* as this element is precipitated at the roots as insoluble calcium vanadate.

20-33   ZINC

Concentration in igneous rocks: 80 ppm
Concentration in soils: 50 ppm
Concentration in plant ash: 1400 ppm
Toxicity to plants: moderate
Geobotanical references: Jensch (1894); Schwickerath (1931)
Biogeochemical references: Basitova et al. (1964) (CA 63, 9672d); Brown and Meyer (1960) (CA 57, 3097f); Chapman and Shacklette (1960) (CA 55, 13191c); Cole et al. (1968); Flerova and Flerov (1964)(CA 62, 14351e); Fortescue and Hornbrook (1969) (CA 71, 93426f); Harbaugh (1950) (CA 45, 716e); Hornbrook (1969) (CA 71, 93458j); Keith (1968); Lyubofeev et al. (1962) (CA 60, 7805e); Malyuga and Makarova (1955) (CA 53, 19251d); Malyuga et al. (1960) (CA 55, 13191h); Nicolas and Brooks (1969); Nicolls et al. (1965) (CA 63, 16048d); Poskotin (1963) (CA 62, 342g); Poskotin

and Lyubimova (1963) (CA 64, 12380f); Riddell (1952); Robinson et al. (1947) (CA 42, 8725d); Salmi (1956) (CA 51, 7256f); Schwickerath (1931); Shacklette (1960) (CA 55, 13736h); Starikov et al. (1964) (CA 61, 15847d); Starr (1949); Talipov (1964b) (CA 62, 6268h), (1965) (CA 64, 12379g); Tyurina and Shchibrik (1962) (CA 58, 1250f); Warren and Delavault (1948) (CA 43, 3321c), (1949) (CA 43, 3321h), (1955b, c); Warren, Delavault, and Cross (1959); Warren Delavault, and Irish (1949) (CA 44, 6787d), (1952a) (CA 46, 6555f); Warren and Howatson (1947) (CA 42, 1533c); Webb and Millman (1951) (CA 45, 9428c); White (1950) (CA 44, 10614c); Worthington (1955) (CA 50, 1536g)

ABSTRACT   Keith (1968). "Relationships of lead and zinc contents of trees and shrubs, Upper Mississippi Valley district." Lead and *zinc* contents of elm, maple, and oak trees, and of soils in which the trees grew, were determined in samples from the Upper Mississippi Valley. Specimens were taken from a mining district and from a nearby background area. The lead and *zinc* analyses of plant materials served to differentiate the mineralized and nonmineralized areas. Stems contained about twice as much lead and two to five times as much *zinc* as did leaves. Lead and *zinc* concentrations in all soil horizons could be used to differentiate between the two areas. Soils at most sampling sites in the mineralized district seemed more reliable than plants for geochemical prospecting and are a more practical sampling medium. Soil samples taken from areas of thick loess cover failed to indicate the presence of mineralized ore bodies, whereas tree samples from the same sites contained anomalous amounts of lead and *zinc*. It is suggested that in areas of thick loess cover, plant samples could be used in prospecting for lead and *zinc*.

APPENDIXES

APPENDIX I   Table of t Distributions

df	$t_{0.60}$	$t_{0.70}$	$t_{0.80}$	$t_{0.90}$	$t_{0.95}$	$t_{0.975}$	$t_{0.99}$	$t_{0.995}$
1	0.325	0.727	1.376	3.078	6.314	12.706	31.821	63.657
2	0.289	0.617	1.061	1.886	2.920	4.303	6.965	9.925
3	0.277	0.584	0.978	1.638	2.353	3.182	4.541	5.841
4	0.271	0.569	0.941	1.533	2.132	2.776	3.747	4.604
5	0.267	0.559	0.920	1.476	2.015	2.571	3.365	4.032
6	0.265	0.553	0.906	1.440	1.943	2.447	3.143	3.707
7	0.263	0.549	0.896	1.415	1.895	2.365	2.998	3.499
8	0.262	0.546	0.889	1.397	1.860	2.306	2.896	3.355
9	0.261	0.543	0.883	1.383	1.833	2.262	2.821	3.250
10	0.260	0.542	0.879	1.372	1.812	2.228	2.764	3.169
11	0.260	0.540	0.876	1.363	1.796	2.201	2.718	3.106
12	0.259	0.539	0.873	1.356	1.782	2.179	2.681	3.055
13	0.259	0.538	0.870	1.350	1.771	2.160	2.650	3.012
14	0.258	0.537	0.868	1.345	1.761	2.145	2.624	2.977
15	0.258	0.536	0.866	1.341	1.753	2.131	2.602	2.947
16	0.258	0.535	0.865	1.337	1.746	2.120	2.583	2.921
17	0.257	0.534	0.863	1.333	1.740	2.110	2.567	2.808
18	0.257	0.534	0.862	1.330	1.734	2.101	2.552	2.878
19	0.257	0.533	0.861	1.328	1.729	2.093	2.539	2.861
20	0.257	0.533	0.860	1.325	1.725	2.086	2.528	2.845
21	0.257	0.532	0.859	1.323	1.721	2.080	2.518	2.831
22	0.256	0.532	0.858	1.321	1.717	2.074	2.508	2.819
23	0.256	0.532	0.858	1.319	1.714	2.069	2.500	2.807
24	0.256	0.531	0.857	1.318	1.711	2.064	2.492	2.797
25	0.256	0.531	0.856	1.316	1.708	2.060	2.485	2.787
26	0.256	0.531	0.856	1.315	1.706	2.056	2.479	2.779
27	0.256	0.531	0.855	1.314	1.703	2.052	2.473	2.771
28	0.256	0.530	0.855	1.313	1.701	2.048	2.467	2.763
29	0.256	0.530	0.854	1.311	1.699	2.045	2.462	2.756
30	0.256	0.530	0.854	1.310	1.697	2.042	2.457	2.750
40	0.255	0.529	0.851	1.303	1.684	2.021	2.423	2.704
60	0.254	0.527	0.848	1.296	1.671	2.000	2.390	2.660
120	0.254	0.526	0.845	1.289	1.658	1.980	2.358	2.617
∞	0.253	0.524	0.842	1.282	1.645	1.960	2.326	2.576

APPENDIX II   Table of Significance Values for r

Degrees of freedom	r				
	0.10	0.05	0.02	0.01	0.001
1	0.988	0.997	0.999	1.000	1.000
2	0.900	0.950	0.980	0.990	0.999
3	0.805	0.878	0.934	0.959	0.992
4	0.729	0.811	0.882	0.917	0.974
5	0.669	0.754	0.833	0.874	0.951
6	0.621	0.707	0.789	0.834	0.925
7	0.582	0.666	0.750	0.798	0.898
8	0.549	0.632	0.716	0.765	0.872
9	0.521	0.602	0.685	0.735	0.847
10	0.497	0.576	0.658	0.708	0.823
11	0.476	0.553	0.634	0.684	0.801
12	0.457	0.532	0.612	0.661	0.780
13	0.441	0.514	0.592	0.641	0.760
14	0.426	0.497	0.574	0.623	0.742
15	0.412	0.482	0.558	0.606	0.725
16	0.400	0.468	0.543	0.590	0.708
17	0.389	0.456	0.528	0.575	0.693
18	0.378	0.444	0.516	0.561	0.679
19	0.369	0.433	0.503	0.549	0.665
20	0.360	0.423	0.492	0.537	0.652
25	0.323	0.381	0.445	0.487	0.597
30	0.296	0.349	0.409	0.449	0.554
35	0.275	0.325	0.381	0.418	0.519
40	0.257	0.304	0.358	0.393	0.490
45	0.243	0.287	0.338	0.372	0.465
50	0.231	0.273	0.322	0.354	0.443
60	0.211	0.250	0.295	0.325	0.408
70	0.195	0.232	0.274	0.302	0.380
80	0.183	0.217	0.256	0.283	0.357
90	0.173	0.205	0.242	0.267	0.337
100	0.164	0.195	0.230	0.254	0.321

APPENDIX III Computer program for calculation of Pearson correlation coefficients (r)

```
C       CORRELATION COEFFICIENTS,GEOMETRIC MEANS,STANDARD DEVIATIONS,
C       REDUCED MAJOR AXES
C       INPUT FORMAT ALLOWS 12 6UNIT FIELDS FOR DATA,
C       1 6UNIT FIELD FOR SAMPLE IDENTIFICATION
C       ALL INPUT DATA LOGARITHMICALLY TRANSFORMED
C       SENSE SWITCHS 1,2 ALLOW MANIPULATION OF INPUT DATA
C       SENSE SWITCH 1 ON FIRST OPTION
C       SENSE SWITCH 2 ON SECOND OPTION
C       TITLE MAY OCCUPY 50 COLUMNS
C       SENSE SWITCH 4 ON READS NEW DATA DECK
C       SENSE SWITCH 3 ON DATA PRINT OUT
        DIMENSION A(13),S(12),SQ(12),AV(12),GM(12),SD(12),B(12)
        DIMENSION SP(66),X(66),Y(66),Z(66),R(66),RMA(66),MD(12)
        DEFINE DISK(13,1000)
    10  READ 20
    20  FORMAT (50H
        IDEX=1
    30  READ 40,A,M
    40  FORMAT (13F6.0,1X,I1)
        IF(M-9)50,60,50
    50  RECORD(IDEX)A
        GO TO 30
    60  ISAVE=IDEX
    70  DO 80 I=1,12
        S(I)=0
        SQ(I)=0
    80  AV(I)=0
        DO 90 J=1,66
    90  SP(J)=0
        AN=0.0
        PRINT 20
C       START OF MAIN LOOP
        IDEX=1
        IF(SENSE SWITCH 1)100,120
   100  PRINT 110
   110  FORMAT (22H FIRST OPTION SUBTITLE)
        GO TO 140
   120  PRINT 130
   130  FORMAT (20H INPUT DATA SUBTITLE)
   140  IF(SENSE SWITCH 2)150,161
   150  PRINT 160
   160  FORMAT (23H SECOND OPTION SUBTITLE)
   161  PRINT 162
   162  FORMAT (//)
   170  FETCH(IDEX) A
        IF(IDEX-ISAVE)180,180,350
   180  IF(SENSE SWITCH 1)190,210
   190  DO 200 I=1,10
C       FIRST OPTION MANIPULATION INSERTED HERE
   200  A(I)=A(I)*A(11)
   210  IF(SENSE SWITCH 2)220,240
   220  DO 230 I=1,12,2
C       SECOND OPTION MANIPULATION INSERTED HERE
   230  A(I)=A(I)/A(I+1)
   240  IF(SENSE SWITCH 3)250,280
   250  PRINT 270,A
C       DATA PRINTOUT FOLLOWS
   270  FORMAT(13F10.4)
   280  AN=AN+1.0
        DO 290 I=1,12
        B(I)=0.43429448*LOGF(A(I))
        S(I)=S(I)+B(I)
   290  SQ(I)=SQ(I)+B(I)*B(I)
        J=0
        I=0
   300  I=I+1
        JJ=12
```

```
    310 J=J+1
        IF(J-66)320,320,330
    320 X(J)=B(I)*B(JJ)
        JJ=JJ-1
        IF(JJ-(I+1))300,310,310
    330 DO 340 J=1,66
    340 SP(J)=SP(J)+X(J)
        GO TO 170
C       END OF MAIN LOOP
C       START OF MAIN CALCULATIONS
    350 DO 360 I=1,12
        AV(I)=S(I)/AN
        GM(I)=EXPF(AV(I)/0.43429448)
    360 SD(I)=SQRTF((SQ(I)-AN*AV(I)*AV(I))/(AN-1.0))
        J=0
        I=0
    370 I=I+1
        KE=12
    380 J=J+1
        IF(J-66)390,390,400
    390 Y(J)=AV(I)*AV(KE)
        Z(J)=(SQ(I)-AN*AV(I)**2)*(SQ(KE)-AN*AV(KE)**2)
        RMA(J)=SD(I)/SD(KE)
        KE=KE-1
        IF(KE-(I+1))370,380,380
    400 DO 410 J=1,66
    410 R(J)=(SP(J)-AN*Y(J))/SQRTF(Z(J))
C       END OF CALCULATIONS
C       PRINT OUT OF RESULTS FOLLOWS
        PRINT 420
    420 FORMAT (/16H GEOMETRIC MEANS)
        PRINT 430,(GM(I),I=1,12)
    430 FORMAT (12F10.4)
        PRINT 440
    440 FORMAT (/12H STD. DEVS. )
        PRINT 450,(SD(I),I=1,12)
    450 FORMAT (12F10.4)
        PRINT 460,AN
    460 FORMAT(/19H NUMBER OF SAMPLES=,F6.0/)
        PRINT 470
    470 FORMAT(25H CORRELATION COEFFICIENTS/)
        KL=1
        DO 480 I=1,12
    480 MD(I)=13-I
    490 PRINT 500,(MD(I),I=1,12)
    500 FORMAT (11H COLUMN NO.,12(2X,F4.0,4X))
        KC=1
        MA=1
        LA=10
    510 NA=MA+LA
        PRINT 520,KC,(R(J),J=MA,NA)
    520 FORMAT (4X,I2,4X,12F10.4)
        KC=KC+1
        LA=LA-1
        MA=NA+1
        IF(KC-12)510,530,530
    530 GO TO (540,580),KL
    540 PRINT 550
    550 FORMAT (/17H SLOPES OF R.M.A./)
        KL=2
    560 DO 570 J=1,66
    570 R(J)=RMA(J)
        GO TO 490
    580 PAUSE
        IF(SENSE SWITCH 4)10,70
C       STANDARD DEVIATIONS IN UNITS OF LOGARITHM TO BASE TEN
        END
```

APPENDIX IV   Computer program for calculation of Spearman rank correlation coefficients (r_s)

```
*ALL STATEMENT MAP
*LIST PRINTER
*NO DIAGNOSTICS
*FANDK0810
C       NAME CBSRANK
C       SPEARMANS RANK CORRELATION
C       DATA IS SUPPLIED IN TWO BLOCKS
C          FIRST HAS CONCENTRATIONS FOR SOILS
C          SECOND HAS CONCENTRATIONS FOR PLANTS
C       FORMAT IS I6 FOR FIRST 7 REGIONS, F6.2 FOR REGION 8 IN PLANT BLOCK
C          THE LAST REGION CONTAINS THE ASH WEIGHT OF THE PLANT (PERCENT)
C       PARAMETERS ARE
C             A - MATRIX OF POINTS
C             N - NO. OF OBS.
C
C          NVA - NO. OF REGIONS TO USE IN 1ST BLOCK
C          NVB - NO. OF REGIONS TO USE IN 2ND BLOCK
C
C          OBJECT IS TO OBTAIN RANK COR. COEFF. BETWEEN ALL ITEMS IN 1ST
C             BLOCK WITH ALL ITEMS IN 2ND BLOCK
C
C       SUBROUTINES REQD....RANK,TIE,SRANK,SPBRF
C
        DIMENSION LOC(24)
        DIMENSION TITLE(10)
        DIMENSION IA(100,8),S(101),P(101),R(202),WT(101)
        COMMON S,P,R
        FLOATF(I)=I
        EQUIVALENCE (S,WT)
        DEFINE DISK(10,2400)
        NR=1
      1 NPASS=0
C       READ A TITLE
        READ 4000,TITLE
C       PRINT TITLE
        PRINT 5000,TITLE
C       READ PARAMETERS REQD.
        READ 1000,N,NVA,NVB
        NV=NVA+NVB
        NVB1=NVB+1
        NN=N+1
        NSEC=NN/10+1
        IDSK=NSEC+1
C       READ DATA OFF CARDS IN TWO BLOCKS AND RECORD ON DISK AFTER RANKING
C       BLOCK 1
        DO 10 J=1,N
        READ 2000,(IA(J,I),I=1,NVA)
     10 CONTINUE
        DO 20 I=1,NVA
        DO 25 J=1,N
        S(J)=IA(J,I)
     25 CONTINUE
        CALL RANK(S,R,N)
        LOC(I)=IDSK
        RECORD(IDSK)(R(L),L=1,NN)
        IDSK=IDSK+1
     20 CONTINUE
C       BLOCK 2
        DO 30 J=1,N
        READ 2000,(IA(J,I),I=1,NVB1)
     30 CONTINUE
        DO 40 I=1,NVB1
        DO 45 J=1,N
        S(J)=IA(J,I)
     45 CONTINUE
        INVB=I+NVA
        LOC(INVB)=IDSK
        IF (I-8)47,46,48
```

```
      46 IDSK3=1
         RECORD(IDSK3)(S(L),L=1,N)
         GO TO 40
      47 RECORD(IDSK)(S(L),L=1,N)
         GO TO 49
      48 RECORD(IDSK)(S(L),L=1,N)
         INVB1=INVB-1
         LOC(INVB1)=LOC(INVB)
      49 IDSK=IDSK+1
      40 CONTINUE
         DO 50 I=1,NVB
         NVAI=NVA+I
         JDSK2=LOC(NVAI)
         FETCH(JDSK2)(S(L),L=1,N)
         CALL RANK(S,R,N)
         INVC=NV+I
         LOC(INVC)=IDSK
         RECORD(IDSK)(R(L),L=1,NN)
         IDSK=IDSK+1
      50 CONTINUE
C
C        DISK NOW CONTAINS
C            1 SOIL DATA (RANKED) STARTING AT NSEC+1
C            2 PLANT DATA (RAW)   STARTING AT LOC(NVA+1)
C            3 PLANT DATA (RANKED) STARTING AT LOC(NV+1)
C            4 WEIGHT PERCENT OF ASH AT IDSK3
C        USING  (1) AND (3) CAN NOW GET RANK CORRELATIONS BETWEEN
C            CONCENTRATION IN THE SOIL AND THE
C            CONCENTRATION IN THE PLANT (ON THE ASH WEIGHT)
C
C        PRINT HEADINGS
         PRINT 3000,NVA,NVB,N
     100 PRINT 3100
         DO 60 J=1,NVA
         JDSK1=LOC(J)
         FETCH(JDSK1)(S(L),L=1,NN)
         DO 65 K=1,NVB
         NVAI=NV+K
         JDSK2=LOC(NVAI)
         FETCH(JDSK2)(P(L),L=1,NN)
         CALL SRANK(S,P,R,N,RS,T,NDF,NR)
         P2=SPRBF(1.,FLOATF(NDF),T**2)
         PRINT 3200,J,K,RS,T,P2,NDF
      65 CONTINUE
      60 CONTINUE
         NPASS=NPASS+1
         IF (NPASS-1)80,90,80
      90 CONTINUE
         PRINT 3300
C
C        CALCULATE CONCENTRATIONS ON A DRY WEIGHT BASIS AND OBTAIN RS
         JDSK2=1
         FETCH(JDSK2)(WT(L),L=1,N)
         DO 70 J=1,NVB
         NVAI=NVA+J
         JDSK2=LOC(NVAI)
         FETCH(JDSK2)(P(L),L=1,N)
         DO 75 I=1,N
         P(I)=P(I)*WT(I)
      75 CONTINUE
         CALL RANK(P,R,N)
         NVAI=NV+J
         JDSK4=LOC(NVAI)
         RECORD(JDSK4)(R(I),I=1,NN)
      70 CONTINUE
C
C        PLANT DATA (CORRECTED FOR ASH WEIGHT),(RANKED) IS NOW STORED ON
C            THE DISK STARTING AT POSITION IDSK4
```

APPENDIX IV (Continued)

```
C
C      NOW GET CORRELATION BETWEEN THE
C          CONCENTRATION IN THE SOIL AND THE
C          CONCENTRATION IN THE PLANT (ON THE DRY WEIGHT)
C
       GO TO 100
   80 CONTINUE
       GO TO 1
 1000 FORMAT(I4,I2,I2)
 2000 FORMAT(12I6)
 3000 FORMAT(32H ***RANK CORRELATION ANALYSIS***/1X,31(1H=)/1X,56HPLANT
      1CONCENTRATIONS EXPRESSED AS FUNCTION OF ASH WEIGHT/36HONUMBER OF V
      2ARIABLES IN SOIL BLOCK =,I5/37HONUMBER OF VARIABLES IN PLANT BLOCK
      3 =,I4/37HONUMBER OF VALUES FOR EACH VARIABLE =,I4)
 3100 FORMAT(40HO  ELEMENTS      RS      T      P(T)     DF/13H ROCK    PLAN
      1T)
 3200 FORMAT(1H ,I4,4X,I4,F6.3,F7.3,F9.4,I7)
 3300 FORMAT(54H1PLANT CONCENTRATIONS EXPRESSED ON BASIS OF DRY WEIGHT)
 4000 FORMAT(10A4)
 5000 FORMAT(1H1,10A4)
       END
```

ÅBERG, B., 1948, The mechanism of toxic action of chlorates and some related substances on young wheat plants [in Swedish]: K. LantbrHögsk. Annlr, v. 15, p. 37–107.

ADRIANI, M. J., 1958, Halophytes [in German], *in* Encyclopedia of plant physiology, v. 4: Berlin, Springer, 1210 p.

AFEROV, Y. A., ZVYAGIN, V. G., ROSLYAKOVA, M. V., ROSLYAKOV, N. A., SHABYNIN, L. L., and EPOV, L., 1968, Gold in rocks, plants and waters of the Darasun deposit [in Russian]: Vopr. Geol. Pribaikal. Zabaikal., no. 3, p. 146–149.

AHRENS, L. H., 1954, The log-normal distribution of the elements: Geochim. et Cosmochim. Acta, v. 5, p. 49–73.

AHRENS, L. H., and TAYLOR, S. R., 1960, Spectrochemical analysis, *in* Methods in geochemistry: New York, Interscience, 464 p.

AHRENS, L. H., and TAYLOR, S. R., 1961, Spectrochemical analysis: Reading, Addison-Wesley, 454 p.

ALDRICH, D. G., and TURRELL, F. M., 1951, Effect of acidification on some chemical properties of a soil and the plants grown thereon: Soil Sci., v. 70, p. 83–90.

ALESKOVSKY, V. B., MOKHOV, A. A., and SPIROV, V. A., 1959, The use of the biogeochemical method for exploration of nickel on the Kola Peninsula [in Russian]: Geokhimiya, no. 3, p. 266–271.

ANANICHEV, A. V., 1960, Biogeochemical aspects of a province in Northeastern Kazakhstan with a high boron content [in Russian]: Trudy biogeokhim. Lab., v. 11, p. 226–231.

ANDERSON, R. Y., and KURTZ, E. B., JR., 1954, Factors influencing uranium accumulation in plants: Geol. Soc. America Bull., v. 65, p. 1226.

ANDERSON, R. Y., and KURTZ, E. B., JR., 1955, Biogeochemical reconnaissance of the Annie Laurie uranium prospect, Santa Cruz County, Arizona: Econ. Geology, v. 50, p. 227–232.

ANDERSON, R. Y., and KURTZ, E. B., JR., 1956, A method for determination of alpha-radioactivity in plants as a tool for uranium prospecting: Econ. Geology, v. 51, p. 64–68.

ANDREWS-JONES, D. A., 1968, The application of geochemical techniques to mineral exploration: Mineral Ind. Bull., v. 11, no. 6, 31 p.

ARIPOVA, K., and TALIPOV, R. M., 1966, Concentration of gold in soils and plants in the southern part of the Tamdytau Mountains [in Russian]: Uzbek. geol. Zhur., v. 10, p. 45–51.

ARMANDS, G., and LANDERGREN, S., 1960, Geochemical prospecting for uranium in northern Sweden: the enrichment of uranium in peat: Copenhagen, Internat. Geol. Cong., 21st, Rept., pt. 15, p. 51–66.

AUBREY, K. V., 1956, Frequency distributions of elements in igneous rocks: Geochim. et Cosmochim. Acta, v. 9, p. 83–89.

BABIČKA, J., 1943, Gold in living organisms [in German]: Mikrochim. Acta, v. 31, p. 201–253.

BAILEY, F. M., 1899, The Queensland flora: Brisbane, Diddams, 2015 p.

BALABANOV, V. I., and KOVALEVSKY, A. L., 1963, Aerial prospecting of uranium ores in wooded areas [in Russian]: Atomn. Energ., v. 15, p. 432–434.

BARANOV, V. I., 1933, Use of symbols in geobotanical maps [in Russian]: Izv. biol. nauchno-issled. Inst. biol. Sta. perm. gosud. Univ., v. 8, p. 6–8.

BARANOV, V. I., and KUNASHEVA, K. G., 1954, Content of radioactive elements of the thorium series in plants [in Russian]: Trudy biogeokhim. Lab., v. 10, p. 94–97.

BARSHAD, I., 1948, Molybdenum content of pasture plants in relation to toxicity in cattle: Soil Sci., v. 66, p. 187–196.

BARSHAD, I., 1951, Factors affecting the molybdenum content of pasture plants; I Nature of the soil molybdenum, growth of plants and soil; II Effect of soluble phosphates, available nitrogen and soluble sulfates: Soil Sci., v. 71, p. 297–398.

BASITOVA, S. M., and ZASORINA, E. F., 1965, Biogeochemistry of strontium [in Russian]: Dokl. Akad. Nauk tadzhik. SSR, v. 8, p. 20–23.

BASITOVA, S. M., ZASORINA, E. F., KHAMIDOVA, R. M., and KALASHNIKOVA, G. M., 1964, Biogeochemical research in Northern Tadzhikistan [in Russian]: Izv. Akad. Nauk tadzhik. SSR, Otd. fiz-tekh. khim. Nauk, no. 1, p. 30–40.

BASOLO, F., and PEARSON, R. G., 1958, Mechanisms of inorganic reactions; a study of metal complexes: New York, Wiley, 426 p.

BAUMEISTER, W., 1954, The influence of zinc in *Silene inflata* Smith [in German]: Ber. dt. bot. Gesell., v. 67, p. 205–213.

BAY, R. R., 1967, Groundwater and vegetation in two peat bogs in Northern Minnesota: Ecology, v. 48, p. 308–310.

BAYER, E., EGETER, H., FINK, A., NETHER, K., and WEGMAN, K., 1966, Complex formation and flower colors: Angew. Chem. int. Edn., v. 5, p. 791–798.

BAZILEVSKAYA, N. A., and SIBIREVA, Z. A., 1950, Change in the color of the corolla in *Eschscholtzia californica* under the influence of microelements [in Russian]: Leningrad, Byull. glavn. bot. Sada, no. 6, p. 32–38.

BEATH, O. A., EPPSON, H. F., and GILBERT, C. S., 1935, Selenium and other toxic minerals in soils and vegetation: Wyoming Agric. Exp. Sta. Bull., v. 206, 55 p.

BEATH, O. A., GILBERT, C. S., and EPPSON, H. F., 1939a, The use of indicator plants in locating seleniferous areas in Western USA. I General: Am. Jour. Botany, v. 26, p. 257–269.

BEATH, O. A., GILBERT, C. S., and EPPSON, H. F., 1939b, The use of indicator plants in locating seleniferous areas in Western USA; II Correlation: Am. Jour. Botany, v. 26, p. 296–315.

BECK, A. C., REED, J. J., and WILLETT, R. W., 1958, Uranium mineralization in the Hawks Crag breccia of the Lower Buller Gorge region, South Island, New Zealand: New Zealand Jour. Geology Geophysics, v. 1, p. 432–450.

BELL, J. M., CLARKE, E. deC., and MARSHALL, P., 1911, The geology of the Dun Mountain Sub-division: New Zealand Geol. Survey Bull., no. 12, 71 p.

BENEDICT, R. C., 1937, Equisetum; *in* Standard cyclopedia of horticulture, v. 1; New York, Macmillan, 1200 p.

BERBENNI, P., 1959, Research on the arsenic content of soils and vegetation in the mountainous region of the Prealpi Orobiche [in Italian]: Annali Chim., v. 49, p. 614–623.

BERTRAND, D., 1950, The biogeochemistry of vanadium, *in* Survey of contemporary knowledge of biogeochemistry: Am. Mus. Nat. Hist. Bull., v. 94, p. 409–455.

BILLINGS, W. D., 1950, Vegetation and plant growth as affected by chemically altered rocks in the Western Great Basin: Ecology, v. 31, p. 62–74.

BLOSS, D., and STEINER, R. L., 1960, Biogeochemical prospecting for manganese in Northeast Tennessee: Geol. Soc. America Bull., v. 71, p. 1053–1066.

BOICHENKO, E. A., SAENKO, G. N., and UDEL'NOVA, T. M., 1968, Evolution of the concentration function of plants in the biosphere [in Russian]: Geokhimiya, no. 10, p. 1260–1264.

BONDAR, W. F., 1969, Some principles of geochemical analysis: Proc. Internat. Geochem. Explor. Symp., 1st, Colorado, p. 19–22.

BOROVIK-ROMANOVA, T. F., 1965, Content of lithium in soils [in Russian]: Probl. Geokhim., 1965, p. 620–626.

BOTOVA, M. M., MALYUGA, D. P., and MOISEENKO, U. I., 1963, Experience with the use of the biogeochemical method for exploration for uranium under desert conditions [in Russian]: Geokhimiya, no. 4, p. 361–369.

BOWEN, H. J. M., 1966, Trace elements in biochemistry: London, Macmillan, 241 p.

BOWEN, H. J. M., and DYMOND, J. A., 1955, Strontium and barium in plants and soils: London, Proc. Royal Soc., v. B144, p. 355–368.

BOWEN, N. L., 1928, The evolution of the igneous rocks: Princeton, Princeton University Press, 332 p.

BOWES, W. A., BOLES, W. E., and HASELTON, G. M., 1957, Geology of the uraniferous bog deposit at Petit Ranch, Kern County, California: U.S. Atomic Energy Comm., RME-2063, pt. 1, 29 p.

BOYLE, R. W., 1958, Geochemical prospecting in permafrost regions of Yukon, Canada: Mexico City, Proc. Geochem. Explor. Symp., Internat. Geol. Cong., 20th, p. 175–188.

BOYLE, R. W., and CRAGG, C. B., 1957, Soil analysis as a method of geochemical prospecting in Keno Hill–Galena Hill area, Yukon Territory: Bull. Mines Brch. Canada, No. 39, 27 p.

BRADIS, A. V., MALINOVSKY, V. S., and SOROKIN, V. K., 1963, Contents of copper, molybdenum, manganese, cobalt, zinc and silver in wild and domestic plants of the Kalinin region [in Russian]: Trudy kalinin, gos. med. Inst., no. 10, p. 19–23.

BRADSHAW, A. D., 1952, Populations of *Agrostis tenuis* resistant to lead and zinc poisoning: London, Nature, v. 169, p. 1098.

BRAUN-BLANQUET, J., 1932, Plant Sociology: New York, McGraw-Hill, 437 p.

BRENCHLEY, W. E., 1936, The essential nature of certain minor elements in plant nutrition: Bot. Rev., v. 2, p. 173–196.

BROOKES, C. J., BETTELEY, I. G., and LOXSTON, C. M., 1966, Mathematics and statistics for chemists: New York, Wiley, 418 p.

BROOKS, R. R., 1960, Dissolution of silicate rocks: London, Nature, v. 185, p. 837.

BROOKS, R. R., 1968, Biogeochemical prospecting in New Zealand: New Zealand Sci. Rev., v. 26, p. 9–12.

BROOKS, R. R., and LYON, G. L., 1966, Biogeochemical prospecting for molybdenum in New Zealand: New Zealand Jour. Sci., v. 9, p. 706–718.

BROOKS, R. R., PRESLEY, B. J., and KAPLAN, I. R., 1967, APDC–MIBK extraction system for the determination of trace elements in saline waters by atomic absorption spectrophotometry: Talanta, v. 14, p. 809–816.

BROOKS, R. R., and WHITEHEAD, N. E., 1968, A fluorimetric attachment for an atomic absorption spectrophotometer: Jour. Scient. Instrum., ser. 2, v. 1, p. 879–881.

BROWN, A. C., 1964, Geochemistry of the Dawson settlement bog manganese deposit, New Brunswick: Geol. Survey Canada Paper, 63–42, 26 p.

BROWN, D., 1954, Methods of surveying and measuring vegetation: Hurley, Commonw. Bur. Past. Fld. Crops, 223 p.

BROWN, J. S., and MEYER, P. A., 1960, Geochemical prospecting as applied by the St. Joseph Lead Co.: Mexico City, Proc. Geochem. Explor. Symp., Internat. Geol. Cong., 20th, p. 623–639.

BRUNDIN, N., 1939, Method of locating metals and minerals in the ground: U.S. Pat. 2158980.

BUCK, L. J., 1951, Shrub aids in determining extent of orebody: New York Bot. Grdn. Jour., Jan.–Feb., p. 22.

BURENKOV, E. K., and KUZINA, K. I., Reliability of plants as indicators in mineral exploration: Internat. Geol. Rev., v. 10, p. 421–425.

BURGHARDT, H., 1956, Iron–manganese antagonisms in plants [in German]: Flora [Jena], v. 143, p. 1–30.

BUYALOV, N. I., and SHVYRYAYEVA, A. M., 1961, Geobotanical method in prospecting for salts of boron: Internat. Geol. Rev., v. 3, p. 619–625.

CABROL, J., CABROL, P., and MAGNY, J., 1961, Cobalt in plants [in French]: Soc. Chim. biol. Bull., v. 43, p. 1111–1120.

CANNEY, F. C., MYERS, A. T., and WARD, F. N., 1957, A truck-mounted spectrographic laboratory for use in geochemical exploration: Econ. Geology, v. 52, p. 289–306.

CANNON, H. L., 1952, The effect of uranium-vanadium deposits on the vegetation of the Colorado Plateau: Am. Jour. Sci., v. 250, p. 735–770.

CANNON, H. L., 1953, Geobotanical reconnaissance near Grants, New Mexico: U.S. Geol. Survey Circ. 264, p. 1–8.

CANNON, H. L., 1955, Geochemical relations of zinc-bearing peat to the Lockport dolomite, Orleans County, New York: U.S. Geol. Survey Bull. 1000–D, p. 119–185.

CANNON, H. L., 1957, Description of indicator plants and methods of botanical prospecting for uranium deposits on the Colorado Plateau: U.S. Geol. Survey Bull. 1030–M, p. 399–516.

CANNON, H. L., 1959, Advances in botanical methods of prospecting for uranium in western United States: Mexico City, Proc. Geochem. explor. Symp., Internat. Geol. Cong., 20th, p. 235–241.

CANNON, H. L., 1960a, The development of botanical methods of prospecting for uranium on the Colorado Plateau: U.S. Geol. Survey Bull. 1085–A, p. 1–50.

CANNON, H. L., 1960b, Botanical prospecting for ore deposits: Science, v. 132, p. 591–598.

CANNON, H. L., 1960c, Geochemistry of sandstones and related vegetation in the Yellow Cat area of the Thompson district, Grand County, Utah: U.S. Geol. Survey Prof. Paper 400–B, p. 96–97.

CANNON, H. L., 1963, The biogeochemistry of vanadium: Soil Sci., v. 96, p. 196–204.

CANNON, H. L., 1964, Geochemistry of rocks and related soils and vegetation in the Yellow Cat area, Grand County, Utah: U.S. Geol. Survey Bull. 1176, 127 p.

CANNON, H. L., and BOWLES, J. M., 1962, Contamination of vegetation by lead tetraethyl: Science, v. 137, 7 September, 1962, p. 765–766.

CANNON, H. L., and KLEINHAMPL, F. J., 1956, Botanical methods of prospecting for uranium: U.S. Geol. Survey Prof. Paper 300, p. 681–686.

CANNON, H. L., SHACKLETTE, H. T., and BASTRON, H., 1968, Metal absorption by *Equisetum* (horsetail): U.S. Geol. Survey Bull. 1278–A, p. 1–21.

CANNON, H. L., and STARRETT, W. H., 1956, Botanical prospecting for uranium on La Ventura Mesa, Sandoval County, New Mexico: U.S. Geol. Survey Bull. 1009–M, p. 391–407.

CARLISLE, D., and CLEVELAND, G. B., 1958, Plants as a guide to mineralization: California Div. Mines Geology Spec. Rept. 50, 31 p.

CASSIE, R. M., 1954, Some uses of probability paper in the analysis of size frequency distributions: Australian Jour. Marsh. Freshw. Resources, v. 5, p. 513–522.

CHAMBERLAIN, G. T., and SEARLE, A. J., 1963, Trace elements in some East African soils and plants; II Manganese: East African Agric. For. Jour., v. 29, p. 114–119.

CHANDLER, R. F., 1942, The time required for podzol profile formation as evidenced by the Mendenhall glacial deposits near Juneau, Alaska: Soil. Sci. Soc. America Proc., v. 7, p. 454–459.

CHAPMAN, R. M., and SHACKLETTE, H. T., 1960, Geochemical exploration in Alaska: U.S. Geol. Survey Prof. Paper 400–B, p. 104–107.

CHARTKO, M. K., 1967, Manganese in soils, waters and some plants in the eastern part of Brest-Polesye [in Russian]: Vestsi Akad. Navuk. BSSR, Ser. sel'sk. Navuk, no. 4, p. 42–46.

CHAYES, F., 1954, The log-normal distribution of the elements (a discussion): Geochim. et Cosmochim. Acta, v. 6, p. 119–120.

CHEBAEVSKAYA, V. S., 1960, Dynamics of the content and distribution of microelements in plants [in Russian]: Izv. sel!-khoz. Akad. K. A. Timiryazeva, no. 5, p. 123–134.

CHIKISHEV, A. G., 1965, Plant indicators of soils, rocks and sub-surface waters: New York, Consultants Bureau, 209 p.

CHIU, JUNG-CH'IN, and WANG, T. C., 1964, A study on a copper mine with microbiological research [in Chinese]: Acta microbiol. sin., v. 10, p. 157–167.

CHOU-CHIN-HAN, 1960, The results and future of scientific research in geophysical prospecting in Communist China: Internat. Geol. Rev., v. 2, p. 361–365.

CLARKE, F. W., 1924, The data of geochemistry: U.S. Geol. Survey Bull. 770, 841 p.

COHEN, N. E., BROOKS, R. R., and REEVES, R. D., 1967, The occurrence of beryllium in the Hawks Crag breccia of the Lower Buller Gorge region of New Zealand: New Zealand Jour. Geology Geophysics, v. 10, p. 732–741.

COHEN, N. E., BROOKS, R. R., and REEVES, R. D., 1969, Pathfinders in geochemical prospecting for uranium in New Zealand: Econ. Geology, v. 64, p. 519–525.

COLE, M. M., PROVAN, D. M. J., and TOOMS, J. S., 1968, Geobotany, biogeochemistry, and geochemistry in the Bulman-Waimuna Springs area, Northern Territory, Australia: Inst. Mining Metallurgy Trans., sec. B, p. 81–104.

COLWELL, R. N., 1964, Aerial photography—a valuable sensor for the scientist: Am. Scient., v. 52, p. 17–49.

COLWELL, R. N., 1967, Remote sensing as a means of determining ecological conditions: Bioscience, v. 17, p. 444–449.

COLWELL, R. N., 1968, Remote sensing of natural resources: Sci. Am., v. 218, p. 54–69.

COLWELL, R. N., et al, 1963, Basic matter and energy relationships involved in remote reconnaissance: Photogramm. Eng., v. 29, p. 761–799.

CORBETT, J. R., 1969, The living soil: Sydney, Martindale Press, 326 p.

CUYLER, R. H., 1931, Vegetation as an indicator of geological conditions: Am. Assoc. Petroleum Geologists Bull., v. 15, p. 67–78.

DEBNAM, A. H., 1955, Biogeochemical investigations in the Northern Territory, 1954: Commonw. Australia Dep. Natl. Devel. Bur. Min. Resour. Geology Geophysics Recs., 1955/43, 24 p.

DEBNAM, A. H., 1969, The modern geochemical laboratory: Proc. Internat. Geochem. Explor. Symp., 1st, Colorado, p. 217–221.

DOBROLYUBSKY, O. K., and VOZDEYSTVY, O., 1964, Effects of traces of tungsten in plants [in Russian]: Nauch. Dokl, uyssh. Shk., Ser. biol. Nauk, no 1, p. 142–145.

DOBROVOL'SKY, V. V., 1963, Distribution of trace elements between the soil-forming ground layer, soil and vegetation under the conditions of the Moscow region [in Russian]: Nauch. Dokl. uyssh. Shk., Ser. biol. Nauk, no. 3, p. 193–198.

DOKUCHAEV, V. V., 1883, The Russian chernozem [in Russian]: St. Petersburg, 376 p.

DORN, P., 1937, Plants as a guide to mineral deposits [in German]: Biologe, v. 6, p. 11–13.

DRAEGER, W. C., and LAUER, D. T., 1967, Present and future forestry applications of remote sensing from space: Meet. Am. Inst. Aeronaut. Astronaut, 4th, Anaheim, Paper 67–765.

DUVIGNEAUD, P., 1958, The vegetation of Katanga and its metalliferous soils [in French]: Soc. r. Bot. Belgique Bull., v. 90, p. 127–286.

DUVIGNEAUD, P., 1959, Plant cobaltophytes in Upper Katanga [in French]: Soc. r. Bot. Belgique Bull., v. 91, p. 111–134.

ECKEL, E. B., WILLIAMS, J. S., GALBRAITH, F. W., et al. (1949), Geology and ore deposits of the La Plata district, Colorado: U.S. Geol. Survey Prof. Paper 219, 179 p.

ELLENBERG, H., 1958, Soil responses [in German]: *in* Encyclopedia of plant physiology, v. 4: Berlin, Springer, 1210 p.

ELLIS, J. A., 1966, Geochemical survey work in the area of the Pariwhakaoho River, near Takaka, Northwest Nelson: New Zealand D.S.I.R. Chem. Div. Rept. CD. 75/12 AJE, 9 p.

ELWELL, W. T., and GIDLEY, J. A. F., 1966, Atomic absorption spectrophotometry: Oxford, Pergamon, 139 p.

ERMENGEN, S. W., 1957, Geochemical prospecting in Chibougamau: Canadian Mining Jour., v. 78, p. 99–104.

ERNST, W., 1967, Bibliography on work on heavy-metal plant communities with the exception of serpentine [in German]: Excerpt. bot. sec. B, v. 8, p. 50–61.

EZDAKOVA, L. A., 1964, Lithium in plants [in Russian]: Bot. Zhur. SSSR, v. 49, p. 1798–1800.

EZDAKOVA, L. A., 1966, Lithium content of plants of the Zeravshan Valley [in Russian]: Bot. Zhur. SSSR, v. 51, p. 727–730.

FAO/UNESCO, 1968, Definition of soil units for the soil map of the world: Rome, FAO/UNESCO, 72 p.

FERRES, H. M., 1951, Influence of light and temperature on nutrient uptake and use, with particular reference to zinc: Spec. Conf. Plut. Anim. Nutr. [Australia] Proc., p. 240–243.

FLEROVA, T. P., and FLEROV, V. E., 1964, Experiment on biogeochemical and geobotanical prospecting in the Dzungarian Ala Tau [in Russian]: Sb. Mater. geol. polez. Iskop. Yuzhn. Kazakh., no. 2, p. 144–152.

FORRESTER, J. D., 1942, A native copper deposit near Jefferson City, Montana: Econ. Geology, v. 37, p. 126–135.

FORTESCUE, J. A. C., and HORNBROOK, E. H. W., 1966, A brief survey of progress made in biogeochemical prospecting research at the Geological Survey of Canada, 1962–1965: Geol. Survey Canada Paper 66–54, p. 111–132.

FORTESCUE, J. A. C., and HORNBROOK, E. H. W., 1967, Progress report on biogeochemical research at the Geological Survey of Canada, 1963–1966: Geol. Survey Canada Paper 67–23, pt. 1, 143 p.

FORTESCUE, J. A. C., and HORNBROOK, E. H. W., 1969, Two quick projects, one at a massive sulfide ore body near Timmins, Ontario, and the other at a copper deposit in Gaspé Park, Quebec, 1969: Geol. Survey Canada Paper 67–23, p. 39–63.

FRASER, D. C., 1961a, A syngenetic copper deposit of recent age: Econ. Geology, v. 56, p. 951–962.

FRASER, D. C., 1961b, Organic sequestration of copper: Econ. Geology, v. 56, p. 1063–1078.

FREY-WYSSLING, A., 1935, The essential elements of plant nutrition [in German]: Naturwissenschaften, v. 23, p. 767–769.

FRITZ, N. L., 1967, Optimum methods for using infrared-sensitive color films: Photogramm. Eng., v. 33, p. 1128–1138.

FROELICH, A. J., and KLEINHAMPL, F. J., 1960, Botanical prospecting for uranium in the Deer Flats area, White Canyon District, San Juan County, Utah: U.S. Geol. Survey Bull. 1085–B, p. 51–84.

FUJIMOTO, C. K., and SHERMAN, D. G., 1948, Behavior of manganese in the soil and the manganese cycle: Soil Sci., v. 66, p. 131–145.

GAMBLE, J. F., 1964, A study of strontium, barium and calcium relations in soils and vegetation: Univ. Microfilms, Ann Arbor, Mich., order no. 64–10914, 116 p.

GARRELS, R. M., 1960, Mineral equilibria at low temperature and pressure: New York, Harper & Row, 254 p.

GATES, F. C., 1949, Field manual of plant ecology: New York, McGraw-Hill, 137 p.

GEILMANN, W., and BRÜNGER, K., 1935, The uptake of germanium by plants [in German]: Biochem. Zeitschr. v. 275, p. 387–395.

GERLOFF, G. C., MOORE, D. D., and CURTIS, J. T., 1964, Mineral content of native plants of Wisconsin: Wisconsin Univ. Agric. Exp. Sta. Research Rpt. 14, 26 p.

GLAZOVSKAYA, M. A., 1964, Biological cycle of elements in various landscape zones of the Urals [in Russian]: Fiz. Khim. Biol. Miner. Pochv. SSSR, Akad. Nauk SSSR, Dokl. mezhdunar. Kongr. Pochv. 8, Bucarest, p. 148–157.

GOLDSCHMIDT, V. M., 1937, The principles of distribution of chemical elements in minerals and rocks: Chem. Soc. Jour., pt. 1, p. 655–673.

GOLDSZTEIN, M., 1957, Geobotanical prospecting for uranium in Esterel [in French]: Soc. fr. Minér. Cristallogr. Bull., v. 80, p. 318–324.

GOODALL, D. W., 1952, Quantitative aspects of plant distribution: Biol. Rev., v. 27, p. 194–245.

GORHAM, E., 1956, The ionic composition of some bog and fen waters in the English Lake District: Jour. Ecology, v. 44, p. 142–152.

GRANGER, F., 1962, Vitruvius on architecture, v. 2: Cambridge, Mass., Heinemann, p. 139–141.

GREEN, J., 1959, Geochemical table of the elements for 1959: Geol. Soc. America Bull. v. 70, p. 1127–1284.

GREIG-SMITH, P., 1964, Quantitative plant ecology: London, Butterworths, 2nd ed., 256 p.

GRIFFITHS, J. C., 1967, Scientific method in analysis of sediments: New York, McGraw-Hill, 508 p.

GRIGORYAN, G. B., 1966, Character of distribution of chemical elements in chestnut soils of the Vokhcha River basin [in Russian]: Izv. sel!-khoz. Nauk Minist. Sel!-Khoz. armyansk. SSR, no. 7, p. 57–64.

GRIMALDI, F. S., MAY, I., and FLETCHER, M. H., 1952, U.S. Geological Survey fluorimetric methods of analysis: U.S. Geol. Survey Circ. 199, 20 p.

GRODZINSKY, A. M., 1959, Natural radioactivity in mosses and lichens [in Russian]: Ukr. biokhim. Zhur., v. 16, p. 30–81.

GRODZINSKY, D. M., and GOLUBKOVA, M. G., 1964, The nature of existing radioactivity of the soil and plants in the USSR [in Russian]: Fiziol.-biokhim. Osobennosti Deistviya yadern Izluchen. Rast., Akad. Nauk Ukr. SSR, Inst. Fiziol. Rast., p. 135–146.

GUHA, M., 1961, A study of the trace element uptake of deciduous trees: Ph.D. thesis, University of Aberdeen.

GVOZDANOVIC, S. M., OVERTON, T. R., and SPIERS, F. W., 1967, A study of the concentration of Ra-226 and Th-228 in the living environment: UKAEA Rept. AERE–R5474, p. 175–189.

HANNA, W. J., and GRANT, C. L., 1962, Spectrochemical analysis of the foliage of certain trees and ornamentals for 23 elements: Torrey Bot. Club Bull., v. 89, p. 293–302.

HARBAUGH, D. V., 1950, Biogeochemical investigations in the Tri-State District: Econ. Geology, v. 45, p. 548–567.

HARTMAN, E. L., 1969, The ecology of the "copper moss" *Mielichhoferia mielichhoferi* in Colorado: Bryologist, v. 72, p. 56–59.

HARTSOCK, L., and PIERCE, A. P., 1952, Geochemical and mineralogical methods of prospecting for mineral deposits: U.S. Geol. Survey Circ. 127, 37 p.

HAWKES, H. E., and SALMON, M. L., 1960, Trace elements in organic soil as a guide to copper ore: Internat. Geol. Cong. 21st, Copenhagen, Rept., pt. 2, p. 38–43.

HAWKES, H. E., and WEBB, J. S., 1962, Geochemistry in mineral exploration: New York, Harper & Row, 415 p.

HEDSTRÖM, H., and NORDSTRÖM, A., 1945, New aspects of ore prospecting [in Swedish]: Jernkont. Gruvbyrå Meddn., v. 39, p. 1–44.

HEINSELMAN, M. L., 1963, Forest sites, bog processes and peatland types in the glacial Lake Agassiz region, Minnesota: Ecol. Monogr., v. 33, p. 327–374.

HEINSELMAN, M. L., 1965, String bogs and other patterned organic terrain near Seney, Upper Michigan: Ecol. Monogr., v. 46, p. 185–188.

HENWOOD, W. J., 1857, Notice of the copper turf of Merioneth: Edinburgh New Phil. Jour., v. 5, p. 61–63.

HERDAN, G., 1948, Quality control by statistical methods: London, Nelson, 251 p.

HEWITT, E. J., 1953, Metal inter-relationships in plant nutrition: Jour. Exptl. Botany, v. 4, p. 59–64.

HEWITT, E. J., and NICHOLAS, D. J. D., 1963, Cations and anions: inhibitions and interactions in metabolism and in enzyme activity, *in* Metabolic inhibitors: New York, Academic Press, 753 p.

HINCHEN, J. D., 1969, Practical statistics for chemical research: London, Methuen, 116 p.

HOLMAN, R. H. C., and DURHAM, C. C., 1966, A mobile spectrographic laboratory: Geol. Survey Canada Paper 66–35, 15 p.

HORIZON, 1959, A flower that led to a copper discovery, Ndola, Zambia: RST Group, 1959, v. 1, p. 35–39.

HORN, M. J., and JONES, D. B., 1941, Isolation from *Astragalus pectinatus* of a crystalline amino acid complex containing selenium and sulphur: Jour. Biol. Chem., v. 139, p. 649–660.

HORNBROOK, E. H. W., 1969, Pilot project at Silvermine lead deposit, Cape Breton Island, Nova Scotia: Canadian Geol. Survey Canada Paper 67–23, pt. 2, p. 65–94.

HORVATH, E., 1960, Uranium adsorption on peat in natural waters containing uranium traces [in Hungarian]: Atomki Közl., v. 2, p. 177–183.

HOWARD-WILLIAMS, C., 1970, The ecology of *Becium homblei* in Central Africa with special reference to metalliferous soils: Jour. Ecology, v. 58, p. 745–763.

HUTCHINSON, G. E., 1943, The biogeochemistry of aluminum and of certain related elements: Quart. Rev. Biology, v. 18, p. 1–29, p. 129–153, p. 242–262, and p. 331–363.

HVATUM, O., 1964, Geochemical investigations of some Norwegian bogs [in Norwegian]: Nordic. geol. winter meeting, 6th, Trondheim, Abs.

IGNATIEFF, V., 1952, Efficient use of fertilizers: London, Hill, 158 p.

IGOSHINA, K. N., 1966, Specificity of the flora and vegetation of ultrabasic rocks in the Polar Urals [in Russian]: Bot. Zhur. SSSR, v. 51, p. 322–338.

IL'IN, V. B., 1966, Biogeochemistry of molybdenum in areas of Southwestern Siberia [in Russian]: Izv. Sib. Otd. Akad. Nauk SSSR, Ser. biol.-med. Nauk, no. 3, p. 22–31.

IMBRIE, J., 1956, Biometric methods in the study of invertebrate fossils: Am. Mus. Nat. Hist. Bull., v. 108, p. 211–252.

IRVING, H. M. N. H., and WILLIAMS, R. J. P., 1953, Stability of the complexes of the divalent transition elements: Chem. Soc. Jour., p. 3192–3210.

IVASHOV, P. V., and BARDYUK, V. V., 1967, Biogeochemical investigations at the far eastern tin deposits [in Russian]: Geokhimiya, no. 2, p. 228–232.

IVES, R. L., 1939, Infra-red photography as an aid in ecological surveys: Ecology, v. 20, p. 433–439.

JAMISON, V. C., 1942, Adsorption and fixation of copper in some sandy soils of Central Florida: Soil Sci., v. 53, p. 287–297.

JENNY, H., 1941, Factors of soil formation: New York, McGraw-Hill, 281 p.

JENNY, H., et al., 1951, Minerals useful to California agriculture. Exploring the soils of California: California Div. Mines Bull., no. 155, p. 9–66.

JENSCH, E., 1894, Zinc flora of Upper Silesia [in German]: Zeitschr. angew. Chem., v. 1, p. 14–15.

JOHNSON, C. M., ASHER, C. J., and BROYER, T. C., 1966, Distribution of selenium in plants: Internat. Selenium Biomed. Symp., 1st, Oregon State University, p. 57–75.

KABATA-PENDIAS, A., and BOLIBRZUCH, E., 1964, Molybdenum in soils and plants from the coastal plain [in Polish]: Roczn. Naukro ln., ser. A, v. 88, p. 605–617.

KABIASHVILI, V. I., 1964, Biogeochemical significance of dispersed tungsten [in Russian]: Soobsch. Akad. Nauk gruz. SSR, v. 35, p. 83–86.

KANTOR, M. Z., 1959, Prospecting for boron deposits in Tadzhikistan [in Russian]: Zap. tadzhikist. Otd. vses. miner. Obsch., no. 1, p. 149–157.

KARAMATA, S., 1967, Biogeochemistry of chromium [in French]: Yougoslavia scient. Cons. Acads. RPF Bull., sect. A, v. 12, p. 244–247.

KARASIK, M. A., 1962, Geochemical profiling as a prospecting method for mercury and polymetallic deposits [in Russian]: Byull. nauchno.-tekhn. Inf. Minist. Geol. Okhr. Nedr SSSR, no. 1, p. 60–64.

KARPINSKY, A. M., 1841, Can living plants be indicators of rocks and formations on which they grow and does their occurrence merit the particular attention of the specialist in structural geology? [in Russian]: Zhur. Sadovodstva, nos. 3 and 4.

KASHLEV, V. F., KURBATOV, A. I., and KUZNETSOV, A. V., 1966, Contents of trace elements in tree and bush plants [in Russian]: Dokl. sel!-khoz. Akad. K. A. Timiryazeva, v. 119, p. 269–273.

KASYNOVA, M. S., 1961, Aerovisual geobotanical observations in deserts and semi-arid regions: Internat. Geol. Rev., v. 3, p. 626–628.

KAZITSYN, Y. V., and ALEKSANDROV, G. V., 1960, Concentration of metals by plants over a molybdenum ore deposit under perennial frozen conditions [in Russian]: Mater. vses. nauchno-issled. geol. Inst., no. 32, p. 127–134.

KEITH, J. R., 1968, Relationships of lead and zinc contents of trees and soils, Upper Mississippi Valley District: Prepr. Soc. Mining Eng. AIME, no. 68–L–320, 12 p.

KELLER, W. D., and FREDERICKSON, A. F., 1952, Role of plants and colloidal acids in the mechanism of weathering: Am. Jour. Sci., v. 250, p. 594–608.

KELLEY, A. P., 1923, Soil acidity, an ecological factor: Soil Sci., v. 16, p. 41–54.

KHAMRABAEV, I. K., and TALIPOV, R. M., 1960, Certain results of biogeochemical investigations in Western Kazakhstan [in Russian]: Uzbek. geol. Zhur., no. 5, p. 3–11.

KLEINHAMPL, F. J., 1962, Botanical prospecting for uranium on South Elk Ridge, San Juan County, Utah: U.S. Geol. Survey Bull. 1085–D, p. 105–188.

KLEINHAMPL, F. J., and KOTEFF, C., 1960, Botanical prospecting for uranium in the Circle Cliffs area, Garfield County, Utah: U.S. Geol. Survey Bull. 1085–C, p. 85–104.

KOCHENOV, A. V., ZINEV'YEV, V. V., and LOVALEVA, S. S., 1965, Some features of the accumulation of uranium in peat bogs: Geochemistry, no. 2, p. 65–70.

KOIRTYOHANN, S. R., 1967, Recent developments in atomic absorption and flame emission spectroscopy: Atom. Absorp. Newsl., v. 6, p. 77–84.

KONSTANTINOV, V. M., 1963, Feasibility of using a biogeochemical survey in prospecting for uranium in arid territories [in Russian]: Sov. Geol., v. 6, p. 151–155.

KOVACS, L., 1966, Comparative soil and plant analysis in dolomite and limestone rock swards [in Hungarian]: Bot. Közl., v. 53, p. 175–184.

KOVALEVSKY, A. L., 1962a, Naturally occurring radioactive elements in plants [in Russian]: Izv. sib. Otd. Akad. Nauk SSSR, no. 4, p. 108–114.

KOVALEVSKY, A. L., 1962b, The biogenic accumulation of elements in soils [in Russian]: Izv. sib. Otd. Akad. Nauk SSSR, no. 9, p. 112–115.

KOVALEVSKY, A. L., 1963, Some generalizations on the accumulation by plants of elements of the 2nd group of D. I. Mendeleev's periodic table [in Russian]: Izv. sib. Otd. Akad. Nauk SSSR, Ser. biol.-med. Nauk, 1963, p. 53–61.

KOVALEVSKY, A. L., 1964, Geochemistry of radioactive elements in soils [in Russian]: Trudy Konf. Pochv. sib. dal'n Vost, Novosibirsk, p. 53–60.

KOVALEVSKY, A. L., 1965, Absorption of radium from various kinds of plants [in Russian]: Izv. sib. Otd. Akad. Nauk SSSR, Ser. biol.-med. Nauk, no. 1, p. 43–47.

KOVALEVSKY, A. L., 1966a, Natural radioactive elements of plants of Siberia [in Russian]: Ulan-Ude, Buryat Kn. Izd., 96 p.

KOVALEVSKY, A. L., 1966b, Biogeochemistry of tungsten in plants [in Russian]: Geokhimiya, no. 6, p. 737–744.

KOVAL'SKY, V. V., ANANICHEV, A. V., and SHAKHOVA, I. K., 1965, Boron biogeochemical province of Northwestern Kazakhstan [in Russian]: Agrokhimiya, no. 11, p. 153–169.

KOVAL'SKY, V. V., BASITOVA, S. M., and ZASORINA, E. F., 1965, Biochemistry of strontium and calcium [in Russian], *in* Microelements in agriculture: Tashkent, Uzbek. Akad. Nauk, p. 327–332.

KOVAL'SKY, V. V., BLOKHINA, R. I., ZASORINA, E. F., and NIKITINA, I., 1968, Strontium biogeochemical provinces of Tadzhikistan [in Russian]: Trudy biogeokhim. Lab., v. 12, p. 123–203.

KOVAL'SKY, V. V., and ERMAKOV, V. V., 1967, Tuva biogeochemical province enriched in selenium [in Russian]: Geokhimiya, no. 1, p. 86–97.

KOVAL'SKY, V. V., and PETRUNINA, N. S., 1964, Geochemical ecology and evolutional variation of plants [in Russian]: Dokl. Akad. Nauk SSSR, v. 159, p. 1175–1178.

KOVAL'SKY, V. V., and VORONITSKAYA, E., 1966, The biogenic migration of uranium [in Russian]: Ukr. biokhim. Zhur., v. 38, p. 419–424.

KOVAL'SKY, V. V., and YAROVAYA, G. A., 1966, Biogeochemical provinces rich in molybdenum [in Russian]: Agrokhimiya, no. 4, p. 78–88.

KOVAL'SKY, V. V., and ZASORINA, E. F., 1965, Biogeochemical aspect of strontium [in Russian]: Agrokhimiya, no. 4, p. 78–88.

KRANZ, R. L., 1968, Participation of organic compounds in the transport of ore metals in hydrothermal solution: Inst. Mining Metallurgy Trans., v. 77, p. 26–36.

KRAUSE, W., 1958, Other characteristic floras [in German]; *in* Encyclopedia of plant physiology, v. 4: Berlin, Springer, 1210 p.

KRAUSKOPF, K. B., 1967, Introduction to geochemistry: New York, McGraw-Hill, 721 p.

KRUCKEBERG, A. R., 1954, The ecology of serpentine soils; III Plant species in relation to serpentine soils: Ecology, v. 35, p. 267–274.

KUNASHEVA, K. G., 1944, Radium content in plants and animal organisms [in French]: Trudy biogeokhim Lab., v. 7, p. 98–105.

KUZIN, M. F., 1959, Biogeochemical method for prospecting for rare metal deposits [in Russian]: Razv. Okhr. Nedr, no. 11, p. 16–20.

LAKIN, H. W., ALMOND, H., and WARD, F. N., 1952, Compilation of field methods in geochemical prospecting by the U.S. Geological Survey: U.S. Geol. Survey Circ. 161, 34 p.

LECOQ, J. J., BIGOTTE, G., HINAULT, J., and LECONTE, J. R., 1958, Prospecting for uranium and thorium minerals in desert countries and in the equatorial forest regions of the French Union: Geneva, Proc. Internat. Conf. Peaceful Uses Atom. Energy, 2nd, v. 2, p. 744–786.

LE JOLIS, A., 1860, The chemical influence of the soil on the distribution of plants [in French]: French Cong. Sci., 21st Session, Cherbourg, Procès-verbaux, 38 p.

LEROY, L. W., and KOKSOY, M., 1962, The lichen—a possible plant medium for mineral exploration: Econ. Geology, v. 57, p. 107–113.

LEUTWEIN, F., and PFEIFFER, L., 1954, The possible usefulness of geochemical prospecting for hydrous nickel silicate ores [in German]: Geologie, v. 3, p. 950–1008.

LEUTWEIN, F., and STARKE, R., 1957, Possibility of geochemical selenium prospecting, investigated by the example of the copper schist ore district of Tilkerode [in German]: Geologie, v. 6, p. 349–378.

LEVANIDOV, L. Y., and KHILYUKOVA, M. I., 1953, The problem of the migration of manganese in the weathered crust and in the biosphere in the Southern Urals [in Russian]: Metod Sb. chelyabinsk. gos. ped. Inst., 26 p.

LEWIS, L. L., 1968, Atomic absorption spectrometry—application and problems: Anal. Chem., v. 40, p. 28–47.

LIDGEY, E., 1897, Some indications of ore deposits: Trans. Australas. Inst. Mining Engrs., v. 4, p. 110–122.

LIEBHAFSKY, H. A., PFEIFFER, H. G., WINSLOW, E. H., and ZEMANY, P. D., 1960, X-ray absorption and emission in analytical chemistry: New York, Wiley, 357 p.

LIEBSCHER, K., and SMITH, H., 1968, Essential and nonessential elements: Archs. Envir. Hlth, v. 17, p. 881–890.

LINDNER, M., and SAROSIEK, J., 1963, Biogeochemical examinations in the Sudetes [in Polish]: Przegl. Geol., v. 11, p. 448–450.

LINSTOW, O. VON, 1929, Plant indicators of soil [in German]: Berlin, Preuss. geol. Landesanstalt, 246 p.

LISITSIN, A. K., KRUGLOV, A. I., PANTELEEV, V. M., and SIDEL'NIKOVA, V. D., 1967, Conditions of uranium accumulation in low-lying peat deposits [in Russian]: Litol. polez. Iskop., no. 3, p. 103–116.

LIWSKI, S., 1961, Trace elements manganese, iron, boron, copper, zinc and molybdenum in meadows and marsh plants [in Polish]: Roczn. Naukro ln., ser. F, v. 75, p. 7–14.

LOBANOV, E. M., and KHATAMOV, S., 1968, Stable and natural radioactive elements in plant ash [in Russian]: Izv. Akad. Nauk tadzhik. SSR, Otd. fiz.-mat., no. 2, p. 3–11.

LOBANOV, E. M., KHATAMOV, S., and KHAMIDOVA, R. V., 1967, Determination of gold in biological and geological objects by neutron activation without a chemical decomposition [in Russian]: Aktiv. Anal. Gorn. Porod. Drugikh. Ob'ektov, 1967, p. 147–157.

LOBANOV, E. M., KHATAMOV, S., and KIST, A. A., 1967, Radiochemical and preliminary extraction of gold in neutron activation analysis from samples of plants and geological materials [in Russian]: Aktiv. Anal. Gorn. Porod. Drugikh. Ob'ektov, 1967, p. 158–166.

LOBANOV, E. M., KHATAMOV, S., and TALIPOV, R. M., 1966, Efficiency of neutron activation logging during biogeochemical prospecting for gold deposits in Central Kazakhstan [in Russian]: Uzbek. geol. Zhur., v. 10, p. 49–54.

LÖHNIS, M. P., 1950, Injury through excess of manganese, *in* Trace elements in plant physiology: Lotsya, v. 3, p. 63–76.

LÖHNIS, M. P., 1951, Manganese toxicity in field and market garden crops: Plnt. Soil, v. 3, p. 193–221.

LOUNAMAA, K. J., 1956, Trace elements in plants growing wild on different rocks in Finland. A semi-quantitative spectrographic survey: Annls. bot. Soc. zool.-bot., fennicae Vanamo, v. 29, p. 1–196.

LOUNAMAA, K. J., 1965, Studies on the content of iron, manganese and zinc in macro-lichens: Annls. bot. fennici, v. 2, p. 127–137.

LOUTIT, M. W., LOUTIT, J. S., and BROOKS, R. R., 1967, Differences in molybdenum uptake by micro-organisms from the rhizosphere of *Raphanus sativus* L. grown in two soils of similar origin: Plnt. Soil, v. 27, p. 335–346.

LOVERING, T. S., 1927, Organic precipitation of metallic copper: U.S. Geol. Survey Bull., 795–C, p. 45–52.

LOVERING, T. S., HUFF, L. C., and ALMOND, H., 1950, Dispersion of copper from the San Manuel copper deposit, Pinal County, Arizona: Econ. Geology, v. 45, p. 493–514.

LUNDBERG, H. T. F., 1941, New techniques in geoexploration: Mining Metallurgy, v. 22, p. 256–257.

LUNDEGÅRDH, H., 1931, Environment and plant development: London, Arnold, 330 p.

LYON, G. L., 1969, Trace elements in New Zealand plants, Ph.D. thesis, Massey University.

LYON, G. L., and BROOKS, R. R., 1969, The trace element content of *Olearia rani* and its application to biogeochemical prospecting: New Zealand Jour. Sci., v. 12, p. 200–206.

LYON, G. L., BROOKS, R. R., and PETERSON, P. J., 1969, Chromium-51 distribution in tissues and extracts of *Leptospermum scoparium*: Planta, v. 88, p. 282–287.

LYON, G. L., BROOKS, R. R., PETERSON, P. J., and BUTLER, G. W., 1968, Trace elements in a New Zealand serpentine flora: Plnt. Soil, v. 29, p. 225–240.

LYON, G. L., BROOKS, R. R., PETERSON, P. J., and BUTLER, G. W., 1970, Some trace elements in plants from serpentine soils: New Zealand Jour. Sci., v. 13, p. 133–139.

LYON, G. L., PETERSON, P. J., and BROOKS, R. R., 1969, Chromium-51 transport in the xylem sap of *Leptospermum scoparium* (manuka): New Zealand Jour. Sci., v. 12, p. 541–545.

LYUBOFEEV, V. N., BALITSKY, V. S., and CHERKASOV, M. I., 1962, A biogeochemical method [in Russian]: Trudy geol. polez. Iskop. Sev. Kavk., no. 2, p. 281–287.

MABBUTT, J. A., LITCHFIELD, W. H., SPECK, N. H., SOFOULIS, J., WILCOX, D. G., ARNOLD, J. M., BROOKFIELD, M., and WRIGHT, R. L., 1963, General report on lands of the Wiluna-Meekatharra area, Western Australia, 1958: Land Res. Ser. C.S.I.R.O. Australia, no. 7, 214 p.

MCCARTHY, J. H., JR., 1959, Accuracy and precision of field methods of trace analysis used in geochemical prospecting by the U.S. Geological Survey: Proc., Geochem. Explor. Symp. Internat. Geol. Cong., 20th, Mexico City, p. 363–375.

MCCRAY, C. W. R., and HURWOOD, I. S., 1963, Selenium in Northwestern Queensland associated with a Main Cretaceous formation: Queensland Jour. Agric. Sci., v. 20, p. 475–498.

MacLIVER, P. C. N., MCDONALD, D., SAMPEY, D., and SULLIVAN, J. V., 1969, Multielement analysis in geochemistry: a six-channel atomic absorption spectrophotometer: Proc. Internat. Geochem. Explor. Symp., 1st, Colorado, p. 511.

MAKAROV, M. S., 1965, Biogeochemical prospecting for uranium deposits in an area of the USSR [in Russian]: Vopr. rud. Geofiz. Minist. Geol. Okhr. Nedr SSSR, no. 5, p. 33–39.

MAKAROVA, A. I., 1960, Biogeochemical investigations on polymetallic deposits [in Russian]: Geokhimiya, no. 7, p. 624–633.

MALYUGA, D. P., 1939, The geochemistry of dispersed nickel. I Distribution of nickel in organisms in the biosphere [in Russian]: Trudy biogeokhim. Lab. Akad. Nauk, no. 5, p. 91–108.

MALYUGA, D. P., 1947, Chemical composition of soils and plants as indicators in prospecting for metals [in Russian]: Izv. Akad. Nauk SSSR, Ser. geograf. geofiz., v. 11, p. 135–138.

MALYUGA, D. P., 1950, Biogeochemical provinces in the Southern Urals [in Russian]: Dokl. Akad. Nauk SSSR, v. 70, p. 257–259.

MALYUGA, D. P., 1951, Experience in the soil-floristic method of exploration in the Trans-Ural steppe province [in Russian]: Dokl. Akad. Nauk SSSR, v. 76, p. 231–233.

MALYUGA, D. P., 1954, Experience with the use of the biogeochemical method for exploration of ore deposits in the Southern Urals [in Russian]: Trudy biogeokhim. Lab., v. 10, p. 28–59.

MALYUGA, D. P., 1958a, An experiment in biogeochemical prospecting for molybdenum in Armenia: Geochemistry, no. 3, p. 314–337.

MALYUGA, D. P., 1958b, Copper and molybdenum content of soils and plants over ore deposits [in Russian], in Trace elements in agriculture: ed. J. Peive, p. 105–113.

MALYUGA, D. P., 1959, Application of the biogeochemical method in prospecting for copper–molybdenum ores [in Russian]: Razv. Okhr. Nedr, v. 25, p. 19–22.

MALYUGA, D. P., 1960, Distribution of copper and molybdenum in the soils, waters and plants of the Kadzharan mining district of the Armenian SSR [in Russian]: Trudy biogeokhim. Lab, v. 11, p. 197–207.

MALYUGA, D. P., 1964, Biogeochemical methods of prospecting: New York, Consultants Bureau Enterprises, 205 p.

MALYUGA, D. P., and MAKAROVA, A. I., 1955, Trace elements in virgin soils and in vegetation [in Russian]: Riga, Trudy vses. Soveshch., 1955, p. 485–495.

MALYUGA, D. P., MALASHKINA, N. S., and MAKAROVA, A. I., 1959, Biogeochemical investigations at Kadzharan, Armenian SSR [in Russian]: Geokhimiya, no. 5, p. 423–430.

MALYUGA, D. P., NADIRADZE, V. R., CHARGEISHVILI, Y. M., and MAKAROVA, A. I., 1960, Biogeochemical method of prospecting in the high-altitude region of the Western Georgian SSR [in Russian]: Geokhimiya, no. 4, p. 330–338.

MALYUGA, D. P., and PETRUNINA, N. S., 1961, Biogeochemical investigations in the Tuva Autonomous Republic [in Russian]: Geokhimiya, no. 3, p. 258–267.

MAMULEA, O., and BURACU, O., 1967, Application of biogeochemical methods in prospecting for uranium deposits [in Rumanian]: Dari Seama sedintel. Repub. Soc. Rom. Commn. geol., v. 52, p. 237–244.

MANSKAYA, S. M., DROZDOVA, T. V., and EMELIANOVA, M. P., 1956, Binding of uranium by humic acids and by melanoidins [in Russian]: Geokhimiya, no. 4, p. 10–23.

MANSKAYA, S. M., DROZDOVA, T. V., and EMELIANOVA, M. P., 1958, Fixation of copper by various forms of natural organic compounds [in Russian]: Pochvovedenie, no. 6, p. 41–48.

MANSKAYA, S. M., DROZDOVA, T. V., and EMELIANOVA, M. P., 1960, Distribution of copper in peats and peat soils of the Belo-russian SSR [in Russian]: Geokhimiya, no. 6, p. 630–643.

MARMO, V., 1953, Biogeochemical investigations in Finland: Econ. Geology, v. 48, p. 211–224.

MÅRTENSSON, O., 1956, Bryophytes of the Torneträsk area, Northern Swedish Lappland: K. svenska VetenskAkad. Avh. Naturskydd., no. 14, 321 p.

MATVEEV, A. V., 1961, Aerial prospecting in wooded areas [in Russian]: Atomn. Energ., v. 11, p. 550–552.

MEHTA, B. V., REDDY, G. R., NAIR, G. K., GANDHI, S. C., NEELKANTAN, V., and REDDY, K. G., 1964, Micronutrient studies on Gujarat soils and plants: Indian Soc. Soil Sci. Jour., v. 72, p. 329–342.

MIDDLETON, G. V., 1963, Statistical inference in geochemistry, *in* Studies in analytical geochemistry: Toronto, Univ. Toronto Press, p. 124–139.

MILLER, C. P., 1961, Plant and soil prospecting for nickel: Am. Inst. Mining Metall. Petrol. Engrs. Trans., v. 220, p. 255–260.

MILLER, J. T., and BYERS, H. G., 1937, Selenium in plants in relation to its occurrence in soils: Jour. Agric. Research, v. 55, p. 58–68.

MILLER, R. L., and GOLDBERG, E. D., 1955, The normal distribution in geochemistry: Geochim. et Cosmochim. Acta, v. 8, p. 53–62.

MILLMAN, A. P., 1957, Biogeochemical investigations in areas of copper-tin mineralization in Southwest England: Geochim. et Cosmochim. Acta, v. 12, p. 85–93.

MINGUZZI, C., and NALDONI, K. M., 1950, Supposed traces of arsenic in wood. Its determination in the wood of some trees [in Italian]: Soc. tosc. Sci. nat. Memorie, v. 57, ser. A, p. 38–48.

MINGUZZI, C., and VERGNANO, O., 1948, Nickel content of *Alyssum bertolonii* [in Italian]: Soc. tosc. Sci. nat. Atti, v. 55, p. 49–74.

MINGUZZI, C., and VERGNANO, O., 1953, The inorganic element content of plants of the ultrabasic formation of Impruneta (Florence, Italy) [in Italian]: Nuovo G. bot. ital., v. 60, p. 287–319.

MITCHELL, R. L., 1954, Trace elements in Scottish peats: Dublin, Proc. Internat. Peat Symp., sec. B, p. 3–9.

MITCHELL, R. L., 1955, Trace elements, *in* Chemistry of the soil: New York, Reinhold, 373 p.

MITCHELL, R. L., 1964, The spectrochemical analysis of soils, plants and related materials: Commonw. Bur. Soil. Sci. Tech. Commun. 44A, 225 p.

MITSKEVICH, B. F., 1962, Biochemical prospecting in the Ukraine [in Russian]: Byull. nauchno-tekh. Inf. Minist. Geol. Okhr. Nedr SSSR, no. 1, p. 31–33.

MIZUNO, N., 1967a, Chemical characteristics of serpentine soils in Hokkaido. I The difference in contents of nickel and molybdenum in plants and soils [in Japanese]: Hokkaido-ritsu Nogyo Shikensho Shuho, no. 15, p. 48–55.

MIZUNO, N., 1967b, Chemical characteristics of serpentine soils in Hokkaido. II The difference in contents of zinc, copper, manganese and iron in plants and soils [in Japanese]: Hokkaido-ritsu Nogyo Shikensho Shuho, no. 16, p. 1–9.

MOISEENKO, U. I., 1959, Biogeochemical surveys in prospecting for uranium in marshy areas: Geochemistry, no. 1, p. 117–121.

MORAIN, S. A., and SIMONETT, D. S., 1966, Vegetation analysis with radar imagery: Inst. Sci. Technol. Univ. Mich. Rept. 4864–11–X, p. 605–622.

MORRISON, G. H., and SKOGERBOE, R. K., 1965, Physical methods, *in* Trace analysis: New York, Interscience, p. 10–13.

MORTON, F., and GAMS, H., 1925, Cave plants [in German]: Vienna, Verlag Eduard Hölzel, 227 p.

MULDER, D., 1953, Minor elements in fruit growing [in Italian]: Natl. Conf. Fruitgrowing, Montana di St. Vincent.

MULDER, E. G., and GERRETSEN, F. C., 1952, Soil manganese in relation to plant growth: Adv. Agron., v. 4, p. 222–272.

MURAKAMI, Y., FUJIWARA, S., SATO, M., and OHASHI, S., 1958, Chemical prospecting of uranium deposits in Japan: Geneva, Proc. Internat. Conf. Peaceful Uses Atom. Energy, 2nd, v. 2, p. 131–139.

MUZALEVA, L. D., and PERSHINA, E. F., 1965, Trace element levels in Poaceae and Fabaceae plants in the Suoyarvi district of the Karelian Autonomous SSR [in Russian]: Uchen. Zap. Petrozavodsk. gos. Univ., v. 13, p. 21–28.

MYERS, A. T., and HAMILTON, J. C., 1960, Rhenium in plant samples from the Colorado Plateau: Geol. Soc. America Bull., v. 71, p. 1934.

MYERS, A. T., and HAMILTON, J. C., 1961, Rhenium in plant samples from the Colorado Plateau: U.S. Geol. Survey Prof. Paper 424B, p. 286–288.

NEMEC, B., BABIČKA, J., and OBORSKY, A., 1936, The occurrence of gold in horsetails [in German]: Int. Acad. boheme Sci. Bull. 1–7, 13 p.

NEMODRUK, A. A., and VORONITSKAYA, E., 1962, An extraction luminescent method for determining uranium in soils, silts, plants and animal tissues [in Russian]: Zhur. analit. Khim., v. 17, p. 481–485.

NESVETAYLOVA, N. G., 1961, Geobotanical investigations for prospecting for ore deposits: Internat. Geol. Rev., v. 3, p. 609–618.

NEUMANN, J. VON, and MORGENSTERN, O., 1953, Theory of games and economic behavior: Princeton, Princeton University Press, 641 p.

NICOLAS, D. J., and BROOKS, R. R., 1969, Biogeochemical prospecting for zinc and lead in the Te Aroha region of New Zealand: Australasian Inst. Mining Metall. Proc. 231, p. 59–66.

NICOLLS, O. W., PROVAN, D. M. J., COLE, M. M., and TOOMS, J. S., 1965, Geobotany and biogeochemistry in mineral exploration in the Dugald River area, Cloncurry district, Australia: Inst. Mining Metallurgy Trans., v. 74, p. 696–799.

NIKIFOROV, N. A., and FEDORCHUK, V. P., 1959, Ore indicators in prospecting for concealed ore deposits [in Russian]: Trudy sred.-aziat. politekh. Inst., no. 6, p. 178–190.

NIKONOVA, N. N., 1967, Storage of beryllium, molybdenum, zirconium, yttrium and other rare earth elements in vegetation in the Southwestern Urals [in Russian]: Izv. sib. Otd. Akad. Nauk SSSR, Ser. biol.-med. Nauk, v. 3, p. 25–29.

NOGUCHI, A., 1956, On some mosses of *Merceya* with special reference to the variation and ecology: Kumamoto Jour. Sci., ser. B, v. 2, p. 239–257.

NORMAN, G. G., and FRITZ, N. L., 1965, Infrared photography as an indicator of disease and decline in citrus trees: Florida State Hort. Soc. Proc., v. 78, p. 59–63.

NORTHCOTE, K. H., 1965, A factual key for the recognition of Australian soils: Div. Soils C.S.I.R.O. Australia Rept. 2/65, 112 p.

OHASHI, S., and MURAKAMI, Y., 1960, Progress in geochemical prospecting methods [in Japanese]: Genshiryoku Kogyo, v. 6, p. 41–47.

OSTLE, D., 1954, Geochemical prospecting for uranium: Min. Mag. London, v. 91, p. 201–208.

PAINTER, L. I., TOTH, S. J., and BEAR, F. E., 1953, Nickel status of New Jersey soils: Soil Sci., v. 76, p. 421–429.

PARFENT'EVA, N. S., 1955, Possibility of using vegetation in the exploration for lead deposits in calcareous rocks of the Central Karatau Range [in Russian]: Publication of Moscow State University.

PARIBOK, T. A., and ALEKSEEVA-POPOVA, N. V., 1966, Content of some chemical elements in the wild plants of the Polar Urals as related to the problem of serpentine vegetation [in Russian]: Bot. Zhur. SSSR, v. 31, p. 339–353.

PASSOW, H., ROTHSTEIN, A., and CLARKSON, T. W., 1961, The general pharmacology of the heavy metals: Pharmac. Rev., v. 13, p. 185–224.

PAUL, P. F. M., 1953, Germanium in contemporary vegetation: Notes of a discussion held with Dr. H. Brauchli, Johns Hopkins Univ., Baltimore, 3 p.

PAULI, W., 1968, Some recent developments in biogeochemical research: Canadian Geol. Assoc. Proc., v. 19, p. 45–49.

PEASE, R. W., and BOWDEN, L. W., 1969, Making color infrared film a more effective high-altitude remote sensor: Rem. Sensing Environ., v. 1, 23–30.

PEREL'MAN, A. I., 1967, Geochemistry of epigenesis: New York, Plenum Press, 266 p.

PERSSON, H., 1948, On the discovery of *Merceya ligulata* in the Azores, with a discussion of the so-called "copper mosses": Revue bryol. lichen., v. 17, p. 76–78.

PERSSON, H., 1956, Studies of the so-called "copper mosses": J. Hattori bot. Lab., v. 17, p. 1–18.

PETERSON, P. J., and BUTLER, G. W., 1962, The uptake and assimilation of selenite by higher plants: Australian Jour. Biol. Sci., v. 15, p. 126–146.

PETERSON, P. J., and BUTLER, G. W., 1967, Significance of selenocystathionine in an Australian selenium-accumulating plant, *Neptunia amplexicaulis*: Nature, v. 213, p. 599–600.

PIERCE, R. S., 1953, Oxidation–reduction potential and specific conductance of ground water; their influence on natural forest distribution: Soil Sci. Soc. Am. Proc., v. 17, p. 61–65.

PIRSCHLE, K., 1930, Physiological action of ions [in German]: Jahrb. Wiss. Bot., v. 72, p. 335–368.

PIRSCHLE, K., 1932, Physiological action of homologous ionic series [in German]: Jahrb. Wiss. Bot., v. 76, p. 1–92.

PITKIN, J. A., NEUSCHEL, S. K., and BATES, R. G., 1964, Aeroradioactivity surveys and geologic mapping, *in* The natural radiation environment: Chicago, University of Chicago Press, p. 723–736.

PODKORYTOV, F. M., 1967, Manganese content of plants in the Taimyr region [in Russian]: Trudy nauchno-issled. Inst. sel'.-khoz. krain. Sev., v. 14, p. 277–282.

POLIKARPOCHKINA, V. V., POLIKARPOCHKINA, R. T., and ABRAMOV, I. I., 1965, Dispersion halos in plants in areas of the Eastern Transbaikal ore deposits [in Russian]: Vopr. Geokhim. izverzh. gorn. rudn. Mestorozhd. vost. sib. Akad. Nauk SSSR, sib. Otd., Inst. Geokhim., 1965, p. 242–270.

POLUZEROV, N. A., 1965, The relation of the concentrations of lead, zinc, boron, molybdenum, silver, manganese, and titanium in plants, native rocks, and grey soils [in Russian]: Izv. Akad. Nauk Kazakh. SSR, Ser. biol. Nauk, v. 3, p. 28–37.

POPOV, M. G., 1949, Endemic species of the Maguntan mud volcano [in Russian]: Bot. Zhur. SSSR, v. 34, p. 486–492.

PORUTSKY, G. V., GOLOVCHENKO, V. P., and CHEREDNICHENKO, S. V., 1962, Content of trace elements in various plant organs [in Russian]: Dokl. Akad. Nauk SSSR, v. 146, p. 1223–1226.

POSKOTIN, D. L., 1963, Use of metallometric and biogeochemical methods in prospecting for copper-pyrite deposits in the Middle Urals [in Russian]: Trudy sverdlovsk. gorn. Inst., v. 42, p. 81–94.

POSKOTIN, D. L., and LYUBIMOVA, M. V., 1963, Biogeochemical prospecting for copper sulfide deposits: Geochemistry, no. 6, p. 620–629.

PRAT, S., 1934, The heredity factor of copper tolerance [in German]: Ber. dt. bot. Gesell., v. 52, p. 65–67.

PRATT, P. F., BAIR, F. L., and MCLEAN, G. W., 1967, Nickel and copper chelation capacities of soil organic matter: Internat. Cong. Soil Sci., 8th, Bucharest, Trans., v. 3, p. 243–248.

PUUSTJÄRVI, V., 1956, The cation exchange capacity of various mosses [in Finnish]: Suo, v. 7, p. 53–62.

PUUSTJÄRVI, V., 1968, On the mechanism of uptake of cations by sphagnum [in Finnish]: Suo, v. 9, p. 1–13.

RACKLEY, R. I., SHOCKEY, P. N., and DAHILL, M. P., 1968, Concepts and methods of uranium exploration, in Wyoming Geol. Assoc. Guidebook Ann. Field Conf., v. 20, p. 115–124.

RADFORTH, N. W., 1955, Organic terrain organization from the air (altitudes less than 1000 feet): Canadian Def. Res. Bd. Handbk. DR95, no. 1, 49 p.

RADFORTH, N. W., 1958, Organic terrain organization from the air (altitudes 1000 to 5000 feet): Canadian Def. Res. Bd. Handbk. DR 124, no. 2, 23 p.

RADFORTH, N. W., and USIK, L., 1964, Airphoto interpretation applied to the study of tree growth on bogs: Proc. Muskeg Res. Conf., 9th, NRC Canada, Tech. Memo., no. 81, p. 216–221.

RANKAMA, K., 1940, On the use of trace elements in some problems of practical geology: C. r. Soc. géol. Finl., no. 14, p. 90–106.

RANKAMA, K., 1941, A new method of prospecting [in German]: Geol. Rdsch., v. 32, p. 575–578.

RATSBAUM, E. A., 1939, Field spectroanalytical laboratory for servicing prospecting parties [in Russian]: Razv. Nedr, v. 10, p. 38–41.

RAY, R. G., 1960, Aerial photography in geological interpretation and mapping: U.S. Geol. Survey Prof. Paper 373, 230 p.

RAZIN, L. V., and ROSHKOV, I. S., 1963, Gold geochemistry in the crust of weathering and the biosphere of a permafrost province of the Aldan Shield [in Russian]; Trudy yakutsk. Fil. Akad. Nauk SSSR, Ser. geol., v. 16, p. 5–22.

REICHEN, L. E., 1951, Geochemical field method for determination of nickel in plants: Anal. Chemistry, v. 23, p. 727–729.

REICHEN, L. E., and LAKIN, H. W., 1949, Field method for the determination of zinc in plants: U.S. Geol. Survey Circ. 41, 4 p.

REICHEN, L. E., and WARD, F. N., 1951, Field method for the determination of molybdenum in plants: U.S. Geol. Survey Circ. 124, p. 1–4.

REMEZOV, N. B., 1938, Nitrogen fertilization in pine forests [in Russian]: Sov. Bot., no. 6, p. 34–51.

RIDDELL, J. E., 1952, Anomalous copper and zinc values in trees in Holland township, Gaspé-North County: Quebec Dept. Mines Prelim. Rept. 269, 15 p.

RITCHIE, J. C., 1960, The vegetation of Northern Manitoba. V. Establishing the main zonation: Arctic, v. 13, p. 211–229.

RITCHIE, J. C., 1962, A geobotanical survey of Northern Manitoba: Arctic Inst. N. America Tech. Paper, 9, p. 5–47.

ROBINSON, J. W., 1966, Atomic absorption spectrometry: London, Arnold, 204 p.

ROBINSON, W. O., 1943, The occurrence of rare earths in plants and soils: Soil Sci., v. 56, p. 1–6.

ROBINSON, W. O., 1951, The minor or trace elements in soils, plants and animals: U.S. Dept. Agric. Leaflet, unnumbered, 9 p.

ROBINSON, W. O., BASTRON, H., and MURATA, K. J., 1958, Biogeochemistry of the rare earth elements with particular reference to hickory leaves: Geochim. et Cosmochim. Acta, v. 14, p. 55–67.

ROBINSON, W. O., and EDGINGTON, G., 1942, Boron content of hickory and some other trees: Soil Sci., v. 53, p. 309–319.

ROBINSON, W. O., and EDGINGTON, G., 1945, Minor elements in plants and some accumulator plants: Soil Sci., v. 60, p. 15–28.

ROBINSON, W. O., and EDGINGTON, G., 1948, Toxic aspects of molybdenum in vegetation: Soil Sci., v. 66, p. 197–198.

ROBINSON, W. O., EDGINGTON, G., and BYERS, H. G., 1935, Chemical studies of infertile soils derived from rocks generally high in magnesium and generally high in chromium and nickel: U.S. Dept. Agric. Tech. Bull., 471, 29 p.

ROBINSON, W. O., LAKIN, H. W., and REICHEN, L. E., 1947, The zinc content of plants on the Friedensville zinc slime ponds in relation to biogeochemical prospecting: Econ. Geology, v. 42, p. 572–582.

ROBINSON, W. O., WHETSTONE, R. R., and EDGINGTON, G., 1950, Barium in soils and plants: U.S. Dept. Agric. Tech. Bull., 1013, 36 p.

ROBINSON, W. O., WHETSTONE, R. R., and SCRIBNER, B. F., 1938, Presence of rare earths in hickory leaves: Science, v. 87, p. 470.

ROBYNS, W., 1932, Vegetation and flora of the copper belts of Upper Katanga [in Dutch]: Natuurw. Tijdschr., v. 14, p. 101–106.

ROMMEL, M. A., DAYAN, V. H., SNELL, H., SAUER, A. J., ADBEL-GAWAD, M., and TUFFLY, B. L., 1968, Application of biogeochemistry to mineral prospecting—A survey: NASA Spec. Rept., SP–5056, 134 p.

ROMNEY, E. M., and CHILDRESS, J. D., 1965, Effects of beryllium in plants and soils: Soil Sci., v. 100, p. 210–217.

ROUSE, J. W., WAITE, W. P., and WALTERS, R. L., 1966, Use of orbital radars for geoscience investigations: Lawr. Center Resrch. Eng. Sci., Univ. Kansas, Rept. 61–8, 31 p.

RUNE, O., 1953, Plant life on serpentine and related rocks in the north of Sweden: Acta phytogeogr. suec., v. 31, p 1–135.

SAFRONOV, N. I., POLIKARPOCHKINA, V. V., and UTGOF, A. A., 1958, Spectrometric survey of gold as a method of prospecting for nonplacer gold deposits [in Russian]: Sb. Metod Tekh. geol. Rab., no. 1, p. 100–108.

SALMI, M., 1949, Physical and chemical peat investigations in the Pinimaensuo bog, Southwest Finland: Commn. géol. Finl. Bull., 145, 31 p.

SALMI, M., 1950, On trace elements in peat [in Finnish]: Geotekh. julk. geol. Tutkimuslaitos, 51, 20 p.

SALMI, M., 1955, Prospecting for bog-covered ore by means of peat investigations: Commn. géol. Finl. Bull., 169, p. 5–34.

SALMI, M., 1956, Peat and bog plants as indicators of ore minerals in Vihanti ore field in Western Finland: Commn. géol. Finl. Bull., 175, 22 p.

SALMI, M., 1958, On the pH values of peat as affected by the underlying bedrock [in Finnish]: Geotekh. eripainos geol. Tutkimuslaitos, no. 61, p. 29–39.

SALMI, M., 1959, Peat-chemical prospecting in Finland: Proc. Geochem. Explor. Symp., Internat. Geol. Cong., Mexico City, p. 243–254.

SALMI, M., 1963, On the influence of geological factors upon plant nutrient content of peats [in Finnish]: Maataloust. Aikakausk., v. 35, p. 1–18.

SAMOILOVA, A. P., 1961, Calculation of factors in biogeochemical research [in Russian]: Razv. Okhr. Nedr, v. 27, p. 11–15.

SAMPEY, D., 1969, Newsl. no. 11, 3 p. Pub. Sampey Exploration Services, Perth, Australia.

SANDELL, E. B., 1959, Colorimetric determination of traces of metals: New York, Interscience, 3rd ed., 1032 p.

SANDERS, H. J., 1967, Chemistry of the solid earth: Chem. Eng. News Spec. Rept., p. 2–48.

SAROSIEK, J., 1964, Ecological analysis of some plants growing on serpentine soils in Lower Silesia [in Polish]: Monographiae bot., v. 18, p. 1–105.

SAROSIEK, J., and KLYS, B., 1962, The tin content of plants and soils in Sudety [in Polish]: Soc. Bot. Pol. Acta, v. 31, p. 737–752.

SAUM, N. M., and LINK, J. M., 1969, Exploration for uranium: Mineral Ind. Bull., v. 12, no. 4, 23 p.

SCHATZ, A., 1955, Speculation on the ecology and photosynthesis of the "copper mosses": Bryologist, v. 52, p. 113–120.

SCHOFIELD, W. B., 1959, *Mielichhoferia mielichhoferiana* in the Southern Appalachians: Bryologist, v. 62, p. 248–250.

SCHWANITZ, F., and HAHN, H., 1954a, Genetic-development-physiological investigations on galmei plants [in German]: Zeitschr. Bot., v. 42, p. 179–190.

SCHWANITZ, F., and HAHN, H., 1954b, Genetic-development-physiological investigations on galmei plants II [in German]: Zeitschr. Bot., v. 42, p. 459–471.

SCHWICKERATH, M., 1931, *Viola calaminariae* in zinc soils near Aachen [in German]: Beitr. Nat. Denkm. Pflege, v. 14, p. 463–503.

SELIVANOV, L. S., 1946, Geochemistry and biogeochemistry of dispersed boron [in Russian]: Trudy biogeokhim. Lab., v. 8, p. 5–72.

SERDYUCHENKO, D. P., and PAVLOV, V. A., 1967, Genetic and geochemical characteristics of boron deposits [in Russian]: Redk. Elem. Porod. Razl. metamorf. Fatsii, 1967, p. 126–189.

SHACKLETTE, H. T., 1960, Soil and plant sampling of the Mahoney Creek lead–zinc deposit, Revillagigedo Isl., Southeastern Alaska: U.S. Geol. Survey Prof. Paper 400–B, p. 102–104.

SHACKLETTE, H. T., 1961, Substrate relationships of some bryophyte communities on Latouche Island, Alaska: Bryologist, v. 64, p. 1–16.

SHACKLETTE, H. T., 1962a, Field observations of variations in *Vaccinium uliginosum* L.: Canadian Fld. Natural., v. 76, p. 162–167.

SHACKLETTE, H. T., 1962b, Biotic implications of Alaskan biogeochemical distribution patterns: Ecology, v. 43, p. 138–139.

SHACKLETTE, H. T., 1964, Flower variation of *Epilobium angustifolium* L. growing over uranium deposits: Canadian Fld. Natural., v. 78, p. 32–42.

SHACKLETTE, H. T., 1965a, Element content of bryophytes: U.S. Geol. Survey Bull., 1198–D, 21 p.

SHACKLETTE, H. T., 1965b, Bryophytes associated with mineral deposits and solutions in Alaska: U.S. Geol. Survey Bull., 1198–C, 18 p.

SHACKLETTE, H. T., 1967, Copper mosses as indicators of metal concentrations: U.S. Geol. Survey Bull., 1198–G, 17 p.

SHAKHOVA, I. K., 1960, Biogeochemical aspects of a high-boron province in Northwestern Kazakhstan [in Russian]: Trudy biogeokhim. Lab., no. 11, p. 232–237.

SHAW, D. M., 1961, Element distribution laws in geochemistry: Geochim. et Cosmochim. Acta, v. 23, p. 116–134.

SHAW, W. H. R., 1960, Studies in biogeochemistry. I A biogeochemical periodic table, the data.   II Discussion and references: Geochim. et Cosmochim Acta, v. 19, p. 196–215.

SHCHAPOVA, G. F., 1938, Benthic vegetation of the northeastern bays of the Caspian Sea [in Russian]: Bot. Zhur. SSSR, v. 23, p. 122–143.

SHNYUKOV, E. F., USENK, V. P., and KRASNOZHINA, Z. V., 1963, Results of a biogeochemical survey in the Chivchin manganese deposit [in Russian]: Pit. Geokhim. Miner. Petrogr. Akad. Nauk Ukr. SSR, 1963, p. 107–116.

SHRIFT, A., 1969, Aspects of selenium metabolism in higher plants: Ann. Rev. Plnt. Physiol., v. 20, p. 475–494.

SHRIFT, A., and VIRUPAKSHA, T. K., 1965, Seleno-amino acids in selenium-accumulating plants: Biochim. biophys. Acta, v. 100, p. 65–75.

SHVARTSEV, L. S., 1966, Hydrochemical method of prospecting in northern swamp areas: Internat. Geol. Rev., v. 8., p. 1151–1156.

SHVYRYAYEVA, A. M., and MIKHAILOVA, G. A., 1965, Experiences in the indicator interpretation of geobotanical maps in the North Caspian region, *in*

Plant indicators of soils, rocks and sub-surface waters: New York, Consultants Bureau, 209 p.

SIMMONEAU, P., 1954, Vegetation of saline soils of Oran [in French]: Annls. agronom., sér. A, v. 4, p. 91–117.

SJÖRS, H., 1950a, On the relation between vegetation and electrolytes in north Swedish bog waters [in Swedish]: Oikos, v. 2, p. 241–258.

SJÖRS, H., 1950b, Regional studies in north Swedish bogs [in Swedish]: Bot. Notiser., v. 2, p. 173–222.

SJÖRS, H., 1961, Forest and peatland at Hawley Lake, Northern Ontario; Contributions to botany 1959: Natl. Mus. Canada Bull. 171, p. 1–31.

SJÖRS, H., 1963, Bogs and fens on Attawapiskat River, Northern Ontario; Contributions to botany 1960–1961: Natl. Mus. Canada Bull. 186, p. 45–133.

SKERTCHLY, S. B. J., 1897, The copper plant (*Polycarpea spirostyles*, F.v. Mueller): Queensland Geol. Survey Pub., no. 119, p. 51–53.

SLAVIN, W., 1968, Atomic absorption spectroscopy: New York, Interscience, 307 p.

SMITH, G. D., 1960, Soil classification, a comprehensive system, 7th approximation: Washington, U.S. Dept. Agric., 265 p.

SMITH, G. H., and ELLIS, M. M., 1963, Chromatographic analysis of gases from soils and vegetation, related to geochemical prospecting for petroleum: Am. Assoc. Petroleum Geologists Bull. v. 47, p. 1897–1903.

SOKOLOVA, A. I., and KHRAMOVA, V. V., 1961, Biogeochemical studies [in Russian]: Trudy sverdlovsk. gorn. Inst., no. 40, p. 107–115.

SOKOLOVA, V. Y., and YATSYUK, M. D., 1965, Accumulation of rare and disseminated elements by plants [in Ukrainian]: Ukr. bot. Zhur., v. 22, p. 14–18.

SOMERS, E., 1960, Fungitoxicity of metal ions: Nature [London], v. 187, p. 427–428.

SPARLING, J. H., 1966, Studies on the relationship between water movement and water chemistry in mires: Canadian Jour. Bot., v. 44, p. 747–758.

SPIX, J. B., VON, and MARTIUS, C. F., 1824, Travels in Brazil in the years 1817–1820. Undertaken by command of His Majesty the King of Bavaria: London, Longmans, 2 v. in 1.

STACE, H. C. T., and HUBBLE, G. D., 1968, A handbook of Australian soils: South Australia, Rellim, 435 p.

STANTON, R. E., 1966, Rapid methods of trace analysis: London, Arnold, 96 p.

STARIKOV, V. S., KONOVALOV, B. T., and BRUSHTEIN, I. M., 1964, Biogeochemical method of prospecting and results of its application in the Gornaya Osetia [in Russian]: Geokhimiya, no. 10, p. 1070–1072.

STARR, C. C., 1949, Leaf samples as an aid to prospecting for zinc: West. Miner, v. 22, p. 43.

STEPHENS, C. G., 1962, Manual of Australian soils: Melbourne, C.S.I.R.O., 3rd ed., 61 p.

STILES, W., 1958, Essential micro-(trace) elements; *in* Encyclopedia of plant physiology, v. 4: Berlin, Springer, 1210 p.

STOROZHEVA, M. M., 1954, Teratological phenomena in *Pulsatilla patens* under the conditions of a nickel ore field [in Russian]: Trudy biogeokhim. Lab., v. 10, p. 64–75.

STOUT, P. R., and HOAGLAND, D. R., 1939, Upward and lateral movement of salt in certain plants as indicated by radioactive isotopes of potassium, sodium, and phosphorus absorbed by roots: Am. Jour. Bot., v. 26, p. 320–324.

STUBBS, P., 1968, Spotting uranium from the air: New Scient., v. 37, p. 592–593.

SUTCLIFFE, J. F., 1962, Mineral salt absorption in plants: New York, Pergamon, 194 p.

SZALAY, A., 1954, Investigations of the adsorption of polyvalent cations of high atomic weight on humus colloids [in Hungarian]: Közl. magy. Tudom. Akad., III, v. 4, p. 327–342.

SZALAY, A., 1958, The significance of humus in the geochemical enrichment of uranium: Geneva, Proc. Internat. Conf. Peaceful Uses Atom. Energy, 2nd, v. 2, p. 182–186.

SZALAY, A., and SZILAGYI, M., 1967, The association of vanadium with humic acids: Geochim. et Cosmochim. Acta, v. 31, p. 1–6.

TALIPOV, R. M., 1964a, Concentration of nonferrous metals in soils and plants of the Sary-Cheku and Uch-Kulach deposits (Uzbekistan) [in Russian]: Geokhimiya, no. 5, p. 457–467.

TALIPOV, R. M., 1964b, Results of a biogeochemical survey in some Uzbekistan ore fields. [in Russian]: Polez. Iskop. Uzbek. Vopr. Genezisa, Akad. Nauk uzbek. SSR, 1964, p. 95–102.

TALIPOV, R. M., 1965, Concentration of ore elements in soils and plants of the ore fields in desert and semidesert areas of Uzbekistan [in Russian]: Mikroelemty Sel'sk. Khoz., Akad. Nauk uzbek. SSR, Otd. khim.-tekh. biol. Nauk, 1965, p. 342–353.

TALIPOV, R. M., 1966, Biogeochemical research on polymetallic and copper deposits in Northeastern and Western Uzbekistan [in Russian]: Tashkent, Fan., 104 p.

TALIPOV, R. M., ARIPOVA, K., KARABAEV, K. K., KHATAMOV, S., and AKHUNDKHODZHAEVA, N., 1968, Possible use of arsenic in biogeochemical prospecting for gold ore deposits [in Russian]: Uzbek. geol. Zhur., v. 12, p. 43–47.

TALIPOV, R. M., KHATAMOV, S., KARABAEV, K. K., ARIPOVA, K., and AKHUNKHODZHAEVA, N., 1967, Concentration of rare earths in soils and plants of the Tamdytau area [in Russian]: Uzbek. geol. Zhur., v. 11, p. 17–21.

TCHAKIRIAN, A., 1942, Physiological and therapeutic action of germanium compounds in animals and plants [in French]: Paris, Annls. Inst. Pasteur, v. 68, p. 461–465.

TENNANT, C. B., and WHITE, M. L., 1959, Study of the distribution of some geochemical data: Econ. Geology, v. 54, p. 1281–1290.

THALER, L., 1962, Botanical prospecting as an auxiliary to chemical prospecting for mineral deposits [in French]: Nature, [Paris], no. 3325, p. 208–214.

THORP, J., 1931, The effect of vegetation and climate upon soil profiles in Northern and Northwestern Wyoming: Soil Sci., v. 32, p. 283–301.

THYSSEN, S. W., 1942, Geochemical and geobotanical relationships in the light of applied geophysics [in German]: Beitr. angew. Geophys., v. 10, p. 35–84.

TIMPERLEY, M. H., BROOKS, R. R., and PETERSON, P. J., 1970a, Prospecting for copper and nickel in New Zealand by statistical analysis of biogeochemical data: Econ. Geology, v. 65, p. 505–510.

TIMPERLEY, M. H., BROOKS, R. R., and PETERSON, P. J., 1970b, The significance of essential and nonessential trace elements in plants in relation to biogeochemical prospecting: Jour. Appl. Ecology, v. 7, p. 429–439.

TITAEVA, N. A., 1967, On the character of radium and uranium bonds in peat [in Russian]: Geokhimiya, no. 12, p. 1493–1499.

TKALICH, S. M., 1938, Experience in the investigation of plants as indicators in geological exploration and prospecting [in Russian]: Vest. dal'nevost. Fil. Akad. Nauk SSSR, v. 32, p. 3–25.

TKALICH, S. M., 1952, Botanical methods in geological investigations [in Russian] Bot. Zhur., SSSR, v. 37, p. 660–664.

TKALICH, S. M., 1953, The iron content of plants as an indicator in geological exploration work [in Russian]: Priroda, no. 1, p. 93–95.

TOLGYESI, G., 1962, Trace elements in wild plants [in Hungarian]: Agrokem. Talajt., v. 11, p. 203–218.

TOOMS, J. S., 1959, Field performance of some analytical methods used in geochemical prospecting: Proc. Geochem. Explor. Symp., Internat. Geol. Cong., 20th, Mexico City, p. 377–388.

TOOMS, J. S., and JAY, J. R., 1964, Role of the biochemical cycle in the development of copper/cobalt anomalies in the freely drained soils of the Northern Rhodesian copperbelt: Econ. Geology, v. 59, p. 826–834.

TOWNSEND, J., 1824, Geological and mineralogical researches during a period of more than 50 years in England, Scotland, Ireland, Switzerland, Holland, France, Flanders and Spain: Bath, Gye, 448 p.

TRELEASE, S. F., and BEATH, O. A., 1949, Selenium: its geological occurrence and its biological effects in relation to botany, chemistry, agriculture, nutrition and medicine: New York, Trelease and Beath, 292 p.

TRELEASE, S. F., DiSOMMA, A. A., and JACOBS, A. L., 1960, Seleno-amino acid found in *Astragalus bisulcatus*: Science, v. 132, p. 3427.

TSUNG-SHAN, K., 1957, The skarn type of metamorphic copper ore deposits in Lower Yangtze: Scientia sin., v. 6, p. 1105–1119.

TYUREMNOV, S. N., BARKHATOVA, O. I., and KULIKOVA, G. G., 1968, Trace elements in plants of humid subtropics [in Russian]: Vestn. mosk. gos. Univ., ser. 6, v. 23, p. 47–55.

TYURINA, G. I., and SHCHIBRIK, V. I., 1962, A biogeochemical study of a section of a polymetallic deposit in Central Kazakhstan [in Russian]: Trudy tsent. kazakh. geol. Upr., Minist. Geol. Okhr. Nedr kazakh. SSR, no. 2, p. 44–48.

TYUTINA, N. A., ALESKOVSKY, V. B., and VASIL'EV, P. I., 1959, Practice of biogeochemical prospecting and procedure for determination of niobium in plants [in Russian]: Geokhimiya, no. 6, p. 550–554.

UGLOW, W. L., 1920, Bog manganese deposits, upper north branch Canaan River, Westmorland County, New Brunswick: Munit. Res. Comm. Canada Final Rept. p. 65–79.

USIK, L., 1969, Geochemical and geobotanical prospecting methods in peatland: Geol. Survey Canada Paper 68–66, 43 p.

USPENSKY, Y. Y., 1915, Manganese in plants [in Russian]: Zhur. opyt. Agron, v. 15, p. 425–462.

VAKHROMEEV, G. S., 1962, Biogeochemical exploration methods in the massif of ultrabasic alkaline rocks and carbonatites [in Russian]: Zap. vost.-sib. Otd. vses. miner. Obshch. Akad. Nauk SSSR, no. 4, p. 196–197.

VAUGHN, W. W., 1967, A simple mercury vapor detector for geochemical prospecting: U.S. Geol. Survey Circ. 540, 8 p.

VEKILOVA, F. I., BOROVSKAYA, Y. B., and EFENDIEVA, N. G., 1963, Distribution of cobalt in plants [in Russian]: Izv. Akad. Nauk azerb. SSR, Ser. geol.-geogr. Nauk, no. 4, p. 71–86.

VERKHOVSKAYA, I. N., VAVILOV, P. P., and MASLOV, V. I., 1967, The migration of natural radioactive elements under natural conditions and their distribution according to biotic and abiotic environmental conditions: Internat. Symp. Radioecol. Concn. Processes, Stockholm, Proc. p. 313–328.

VERNADSKY, V. I., 1934, Outline of geochemistry [in Russian]: Moscow, Russkoe Izdaniye, 4th ed., 379 p.

VIKTOROV, S. V., 1955, Use of the geobotanical method in geological and hydrogeological investigations [in Russian]: Moscow, Izv. Akad. Nauk SSSR, 197 p.

VIKTOROV, S. V., 1956, Lichens as indicators of petrographical and geochemical conditions in the desert [in Russian]: Vest. mosk. gos. Univ., no. 5, p. 115–119.

VIKTOROV, S. V., VOSTOKOVA, Y. A., and VYSHIVKIN, D. D., 1962, Introduction to exploration geobotany [in Russian]: Moscow, Moscow Univ., 226 p.

VIKTOROV, S. V., VOSTOKOVA, Y. A., and VYSHIVKIN, D. D., 1964, Short guide to geobotanical surveying: Oxford, Pergamon, 158 p.

VINOGRADOV, A. P., 1959, The geochemistry of rare and dispersed chemical elements in soils: New York, Consultants Bureau, 2nd ed., 209 p.

VINOGRADOV, A. P., 1964, Biogeochemical provinces and their role in organic evolution [in French]: Internat. Monogr. Earth Sci., no. 15, p. 317–337.

VINOGRADOV, B. V., 1963, Development of a photometric method of deciphering aerial photos for automation of vegetation mapping: NASA Rept. 4–28007, p. 47–56.

VIRUPAKSHA, T. K., and SHRIFT, A., 1965, Biochemical differences between selenium-accumulator and non-accumulator *Astragalus* species: Biochim. biophys. Acta, v. 107, p. 69–80.

VISTELIUS, A. B., 1960, The skew frequency distribution and the fundamental law of the geochemical processes: Jour. Geology, v. 68, p. 1–22.

VOGT, T., 1939, Chemical and botanical prospecting at Røros [in Norwegian]: K. norske Vidensk. Selsk. Forh., v. 12, p. 82–83.

VOGT, T., 1942a, Geochemical and geobotanical prospecting. II *Viscaria alpina* G. Don Som "Kisplant" [in Norwegian]: K. norske Vidensk. Selsk. Forh., v. 15, p. 5–8.

VOGT, T., 1942b, Geochemical and geobotanical ore prospecting. III Some notes on the vegetation of the ore deposits at Røros [in Norwegian]: K. norske Vidensk. Selsk. Forh., v. 15, p. 21–24.

VOGT, T., and BERGH, H., 1947, Geochemical and geobotanical prospecting. X Determination of copper in soil samples [in Norwegian]: K. norske Vidensk. Selsk. Forh., v. 19, p. 76–79.

VOGT, T., and BERGH, H., 1948, Geochemical and geobotanical prospecting. XI Zinc and lead in soil samples [in Norwegian]: K. norske Vidensk. Selsk. Forh., v. 20, p. 100–105.

VOGT, T., and BRAADLIE, O., 1942, Geochemical and geobotanical prospecting. IV Vegetation and soils at the Røros ore deposits [in Norwegian]: K. norske Vidensk. Selsk. Forh., v. 15, p. 25–28.

VOGT, T., BRAADLIE, O., and BERGH, H., 1943, Geochemical and geobotanical prospecting. IX Determination of copper, zinc, lead, manganese and iron in plants from Røros [in Norwegian]: K. norske Vidensk. Selsk. Forh., v. 16, p. 55–58.

VOGT, T., and BUGGE, J., 1943, Geochemical and geobotanical prospecting. VIII Determination of copper in plants from Røros with quantitative X-ray analysis [in Norwegian]: K. norske Vidensk. Selsk. Forh., v. 16, p. 51–54.

VOROB'EV, V. Y., and GUDOSHNIKOV, V. V., 1967, Results of a biogeochemical survey on the eastern slope of the Southern Urals [in Russian]: Sov. Geol., v. 10, p. 107–112.

VOSTOKOVA, Y. A., 1957, The botanical method of exploration of uranium-bearing ores [in Russian]: Razv. Okhr. Nedr, v. 23, p. 33–34.

VOSTOKOVA, Y. A., and KHDANOVA, G. I., 1961, Utilization of geobotanical indicators in aeromagnetic mapping in Western Kazakhstan: Internat. Geol. Rev., v. 3, p. 412–416.

VOSTOKOVA, Y. A., VYSHIVKIN, D. D., KASYNOVA, M. S., NESVETAYLOVA, N. G., and SHVYRYAYEVA, A. M., 1961, Geobotanical indicators of bitumen: Internat. Geol. Rev., v. 3, p. 598–608.

WALKER, R. B., 1954, The ecology of serpentine soils. II Factors affecting plant growth on serpentine soils: Ecology, v. 35, p. 259–266.

WALKER, R. B., WALKER, H. M., and ASHWORTH, P. R., 1955, Calcium–magnesium nutrition with especial references to serpentine soils: Plnt. Physiol., v. 30, p. 214–221.

WALLACE, A., and BEAR, F. E., 1949, Influence of potassium and boron on nutrient–element balance in the growth of ranger alfalfa: Plnt. Physiol., v. 24, p. 664–680.

WALLACE, T., 1951, Trace elements in plant physiology: Int. Un. biol. Sci. Colloquia, no. 1, ser. B, p. 5–9.

WALSH, A., 1955, The application of atomic absorption spectra to chemical analysis: Spectrochim. Acta, v. 7, p. 108–117.

WARD, F. N., and NAKAGAWA, H. M., 1969, Atomic absorption techniques in geochemical exploration—problem or progress: Internat. Geochem. Explor. Symp., 1st, Colorado Proc., p. 497–506.

WARINGTON, K., 1937, Observations on the effect of molybdenum in plants with special reference to the Solanaceae: Annls. Appl. Biol., v. 24, p. 475–493.

WARINGTON, K., 1956, Interaction between iron, molybdenum or vanadium in nutrient solutions with or without a growing plant: Annls. Appl. Biol., v. 44, p. 535–546.

WARREN, H. V., 1962, Background data in biogeochemistry: Roy. Soc. Canada Trans., v. 61, p. 21–30.

WARREN, H. V., 1966, Some aspects of lead pollution in perspective: Jour. Coll. Gen Practnrs [Canada], v. 11, p. 135–142.

WARREN, H. V., and DELAVAULT, R. E., 1948, Biogeochemical investigations in British Columbia: Geophysics, v. 13, p. 609–624.

WARREN, H. V., and DELAVAULT, R. E., 1949, Further studies in biogeo-chemistry: Geol. Soc. America Bull., v. 60, p. 531–560.

WARREN, H. V., and DELAVAULT, R. E., 1950a, Gold and silver content of some trees and horsetails in British Columbia: Geol. Soc. American Bull., v. 61, p. 123–128.

WARREN, H. V., and DELAVAULT, R. E., 1950b, History of biogeochemical prospecting in British Columbia: Canadian Inst. Mining Metallurgy Trans., v. 53, p. 236–242.

WARREN, H. V., and DELAVAULT, R. E., 1951, Biogeochemistry and hydro-geochemistry: British Columbia Prof. Engr. April, p. 19–22.

WARREN, H. V., and DELAVAULT, R. E., 1952, Trace elements in geochemistry and biogeochemistry: Scient. Monthly: New York, v. 75, p. 26–30.

WARREN, H. V., and DELAVAULT, R. E., 1955a, Variations in the nickel content of some Canadian trees: Roy. Soc. Canada Trans., v. 48, p. 71–74.

WARREN, H. V., and DELAVAULT, R. E., 1955b, Some biogeochemical investi-gations in Eastern Canada; I and II: Canadian Mining Jour., v. 76, p. 49–54, 58–63.

WARREN, H. V., and DELAVAULT, R. E., 1955c, Biogeochemical prospecting in northern latitudes: Roy. Soc. Canada Trans., v. 49, p. 111–115.

WARREN, H. V., and DELAVAULT, R. E., 1956, Soils in geochemical prospecting: West. Miner, v. 29, p. 36–42.

WARREN, H. V., and DELAVAULT, R. E., 1957, Biogeochemical prospecting for cobalt: Roy. Soc. Canada Trans., v. 51, p. 33–37.

WARREN, H. V., and DELAVAULT, R. E., 1959, Pathfinding elements in geochemical prospecting: Mexico City, Proc. Geochem. Explor. Symp., Internat. Geol. Cong., 20th, p. 255–260.

WARREN, H. V., and DELAVAULT, R. E., 1960, Observations on the biogeochemistry of lead in Canada: Roy. Soc. Canada Trans., v. 54, p. 11–20.

WARREN, H. V., and DELAVAULT, R. E., 1965, Further studies on the biogeochemistry of molybdenum: West. Miner, v. 38, p. 64–68.

WARREN, H. V., DELAVAULT, R. E., and BARAKSO, J., 1964, The role of arsenic as a pathfinder in biogeochemical prospecting: Econ. Geology, v. 59, p. 1381–1389.

WARREN, H. V., DELAVAULT, R. E., and BARAKSO, J., 1966, Some observations on the geochemistry of mercury as applied to prospecting: Econ. Geology, v. 61, p. 1010–1028.

WARREN, H. V., DELAVAULT, R. E., and CROSS, C. H., 1959, Geochemical anomalies related to some British Columbia copper mineralization. Methods and case histories in mining geophysics: Montreal, Proc. Commonw. Mining Metallurgy Cong., 6th, p. 277–282.

WARREN, H. V., DELAVAULT, R. E., and CROSS, C. H., 1966a, Geochemistry in mineral exploration; I: West. Miner, v. 39, p. 22–26.

WARREN, H. V., DELAVAULT, R. E., and CROSS, C. H., 1966b, Geochemistry in mineral exploration; II: West. Miner, v. 39, p. 36–42.

WARREN, H. V., DELAVAULT, R. E., and FORTESCUE, J. A. C., 1955, Sampling in biogeochemistry: Geol. Soc. America Bull., v. 66, p. 229–238.

WARREN, H. V., DELAVAULT, R. E., and IRISH, R. I., 1949, Biogeochemical researches on copper in British Columbia: Roy. Soc. Canada Trans., v. 43, p. 119–137.

WARREN, H. V., DELAVAULT, R. E., and IRISH, R. I., 1951, Further biogeochemical data from the San Manuel copper deposit, Pinal County, Arizona: Geol. Soc. America Bull., v. 62, p. 919–929.

WARREN, H. V., DELAVAULT, R. E., and IRISH, R. I., 1952a, Biogeochemical investigations of the Pacific Northwest: Geol. Soc. America Bull., v. 63, p. 435–484.

WARREN, H. V., DELAVAULT, R. E., and IRISH, R. I., 1952b, Preliminary studies on the biogeochemistry of iron and manganese: Econ. Geology, v. 47, p. 131–145.

WARREN, H. V., DELAVAULT, R. E., and ROUTLEY, D. G., 1953, Preliminary studies of the biogeochemistry of molybdenum: Roy. Soc. Canada Trans., v. 47, p. 71–75.

WARREN, H. V., and HOWATSON, C. H., 1947, Biogeochemical prospecting for copper and zinc: Geol. Soc. America Bull., v. 58, p. 803–820.

WATT, J. S., 1967, The use of gamma-ray excited X-ray sources in X-ray fluorescence analysis: Internat. Jour. Appl. Radiat. Isotopes, v. 18, p. 383–391.

WEAVER, J. E., 1919, The ecological relation of roots: Carnegie Instn. Pub. 286, p. 1–128.

WEBB, J. S., and MILLMAN, A. P., 1951, Heavy metals in vegetation as a guide to ore: Inst. Mining Metallurgy Trans., v. 60, p. 473–504.

WHITE, W. H., 1950, Plant anomalies related to some British Columbia ore deposits: Canadian Inst. Mining Metallurgy Trans., v. 53, p. 368–371.

WHITEHEAD, N. E., and BROOKS, R. R., 1969a, A comparative evaluation of scintillometric, geochemical and biogeochemical methods of prospecting for uranium: Econ. Geology, v. 64, p. 50–55.

WHITEHEAD, N. E., and BROOKS, R. R., 1969b, Radioecological observations on plants of the Lower Buller Gorge region of New Zealand and their significance for biogeochemical prospecting: Jour. Appl. Ecol., v. 6, p. 301–310.

WHITEHEAD, N. E., and BROOKS, R. R., 1969c, Aquatic bryophytes as indicators of uranium mineralization: Bryologist, v. 72, p. 501–507.

WHITEHEAD, N. E., BROOKS, R. R., and COOTE, G. E., 1971, Gamma radiation in some plants and soils from a uraniferous area in New Zealand: New Zealand Jour. Sci., v. 14, p. 66–76.

WHITEHEAD, N. E., BROOKS, R. R., and PETERSON, P. J., 1971, The nature of uranium occurrence in the leaves of *Coprosma australis* (A. Rich.) Robinson: Australian Jour. Biol. Sci., v. 24, p. 67–73.

WHITTAKER, R. H., 1954, The ecology of serpentine soils: Ecology, v. 35, p. 258–259.

WILLIAMS, D. E., and VLAMIS, J., 1957, Manganese toxicity in standard culture solutions: Plnt. Soil, v. 8, p. 183–193.

WILLIAMS, G. J., 1934, The auriferous tin placers of Stewart Island: New Zealand Jour. Sci. Technol., v. 15, p. 344–357.

WILLIAMS, K. T., 1938, Selenium in soils: U.S. Dept. Agric. Yearbk., 1938, p. 831–834.

WILLIAMS, X. K., 1967, Statistics in the interpretation of geochemical data: New Zealand Jour. Geology Geophysics, v. 10, p. 771–797.

WINEFORDNER, J. D., and VICKERS, T. J., 1964, Atomic fluorescence spectrometry as a means of chemical analysis: Anal. Chemistry, v. 36, p. 161–165.

WODZICKI, A., 1959a, Geochemical prospecting for uranium in the Lower Buller Gorge, New Zealand: New Zealand Jour. Geology Geophysics, v. 2, p. 602–612.

WODZICKI, A., 1959b, Radioactive boulders in Hawks Crag breccia: New Zealand Jour. Geology Geophysics, v. 2, p. 385–393.

WORTHINGTON, J. E., 1955, Biogeochemical prospecting of the Shawangunk mine: Econ. Geology, v. 50, p. 420–429.

YAKLOLEVA, M. N., 1964, Fluorimetric (biogeochemical) method for ore prospecting [in Russian]: Byull. nauchno-tekh. Inst. gos. geol. Kom. SSSR, Otd. vses. nauchno-issled. Inst. Min. Syr'ya, no. 3, p. 13–16.

YAROSHENKO, P. D., 1932, On the genesis of mud volcanoes near the Karachal state farm in Southeastern Shirvan [in Russian]: Bot. Sb. azerb. gosud. nauchno-issled. Inst., no. 1.

YAROVAYA, G. A., 1960, Biogeochemical aspects of provinces with increased molybdenum content in the Armenian SSR [in Russian]: Trudy biogeokhim. Lab., no. 2, p. 208–214.

ZAJIC, J. E., 1969, Microbial biogeochemistry: New York, Academic Press, 345 p.

ZAKIROV, K. Z., RISH, M. A., and EZDAKOV, V. I., 1959, Accumulation of microelements in plants growing in regions of ore deposits [in Russian]: Uzbek. biol. Zhur., no. 1, p. 15–20.

ZALASHKOVA, N. E., LIZUNOV, N. V., and SITNIN, A. A., 1958, Beryllium-bearing pegmatites under alluvium and its metallometric survey [in Russian]: Razv. Okhr. Nedr, v. 24, p. 9–14.

INDEXES

AUTHOR INDEX

SUBJECT INDEX

Absorption of ions by plants, 93–94, 96–98
Abundance of trace elements in rocks and soils, 81
Accumulation of elements by plants, 111–112
Aerial gamma measurements, 48–51
Aerial infrared photography, 53, 56, 188
Aerial mapping by satellite, 48
Aerial photography in the ultraviolet and visible ranges, 51–52, 188
Aerial radar imagery, 56–57
Aerial sampling of vegetation, 205
Aerial thermography, 56
Aerovisual mapping, 52
Aerovisual surveys, 52–53
All-Union Aerogeologic Trust, 52
Aluminum
    effect of, on plants, 39
    in plants, 97
    in rocks, 97
    role of, in flower color, 44
Analytical methods
    atomic absorption, 156–158, 161–166, 173–176, 178
    atomic fluorescence, 164, 169–170, 175
    classical, 156
    colorimetry, 156, 158, 164, 172, 175–176
    detection limits of, 164
    emission spectrography, 156–158, 164, 166–168, 173–176, 180
    fluorimetry, 170–171, 175
    neutron activation, 170
    radiometric, 170
    sample preparation in, 157–160
    suitability of, 175
    for volatile substances, 171–172
    X-ray fluorescence, 156–158, 164, 168–169, 175–176
Anomalies in soils, 127, 134–135, 198–199
Anomalies in vegetation, 127, 134–135, 198–199
Antagonistic effects of plant nutrients, 19, 41, 47, 102, 160
Anthocyanins in plant pigments, 44
Antimony
    biogeochemistry of, 209–210
    geobotany of, 209
    in rocks, 209
    in soils, 209
    toxicity of, 210
    in vegetation, 209–210
Aquatic bryophytes in mineral exploration, 177–183

Aqueous migration, coefficient of, 84
Arsenic
    biogeochemistry of, 210
    in rocks, 121–122, 210
    in soils, 121–122, 210
    toxicity of, 210
    in vegetation, 121–122, 210
Arsenopyrite, 121
Autunite, 179
Availability of nutrients to plants, 95, 110

Ballast elements, 96
Barium
    biogeochemistry of, 210–211
    in rocks, 97, 210
    in soils, 210
    toxicity of, 211
    in vegetation, 97, 210
Beryllium
    biogeochemistry of, 211
    in rocks, 211
    in soils, 211
    toxicity of, 211
    in vegetation, 179–180, 211
Biogenic elements, 96–98
Biogeochemical cycle, 88–90
Biogeochemical prospecting
    advantages of, 202
    assessment of, 200–206
    background data in, 127–130
    in British Columbia, 67–69
    case history of, from New Zealand, 190–199
    circumstances in which useful, 123–124
    in Colorado Plateau, 22, 27, 31–32, 68, 106–108
    contamination problems in, 118–121
    in Cornwall, 67
    disadvantages of, 202–203
    economics of, 201–203
    effect of age of plant in, 102–103, 111
    effect of aspect in, 110
    effect of depth of root system in, 106–108
    effect of drainage in, 109–110
    effect of health of plant in, 103, 111
    effect of pH in, 103, 105–106, 111
    effect of plant organ sampled in, 101–102, 111
    effect of type of plant sampled in, 100–101, 111
    element ratios in, 111
    factors of, 204
    factors in choice of, 204–205
    general, 1

INDEX OF BOTANICAL NAMES